Eleventh International Conference on Recirculating Aquaculture 2016

Roanoke, Virginia, USA
19-21 August 2016

ISBN: 978-1-7138-1010-0

Printed from e-media with permission by:

Curran Associates, Inc.
57 Morehouse Lane
Red Hook, NY 12571

RAStech Conferences
PO Box 11736
Blacksburg, VA
24062

https://www.ras-tec.com/

Additional copies of this publication are available from:

Curran Associates, Inc.
57 Morehouse Lane
Red Hook, NY 12571 USA
Phone: 845-758-0400
Fax: 845-758-2633
Email: curran@proceedings.com
Web: www.proceedings.com

TABLE OF CONTENTS

BIOFLOC AND SHRIMP

HATCHERY PRODUCTION TECHNOLOGIES

HEALTH AND DISEASE

POSTERS

RAS SLUDGE MANAGEMENT

SALMONID DIET-NUTRITION FOR RAS

WATER QUALITY CONTROL

2016

11th International Conference on Recirculating Aquaculture

&

Aquaculture Innovation Workshop

Virginia Tech
Invent the Future

The Conservation Fund

August 19-21, 2016
The Hotel Roanoke & Conference Center
Roanoke, Virginia, USA

Welcome

to the

Eleventh International Conference on Recirculating Aquaculture

August 19 - 21, 2016

Hotel Roanoke & Conference Center

Roanoke, Virginia USA

Thursday, August 18, 2016

2:00 p.m. - 8:00 p.m.	Exhibitor Set-up
7:00 p.m. - 10:00 p.m.	Exhibitor's Reception (VIP Suite)

Friday, August 19, 2016

7:30 a.m. – 5:00 p.m.	Registration
7:45 a.m. – 5:00 p.m	Continuous Refreshment Break
8:00 a.m. – 9:00 p.m.	Trade Show Open
8:15 a.m. – 8:30 a.m.	Welcome and Introductory Remarks - Washington Lecture Hall David Kuhn - Virginia Tech Steve Summerfelt - The Conservation Fund's Freshwater Institute
8:30 a.m. - 9:30 a.m.	Keynote John Volpe, Associate Professor, University of Victoria Benchmarking Ecolabeling, and Seafood Watch: How, Why, and What Does It Mean?

	Crystal ABC	Roanoke Ballroom A	Roanoke Ballroom B	Washington
	Aquaponics Moderator: Chris Mullins / Brian Neerie	Advances in Animal Nutrition in RAS Moderator: D. Allen Davis	RAS Production Moderator: Daniel Taylor	Aquaculture Innovation Workshop
9:40 - 10:10	Ryan Chatterson Critical Design Criteria for Engineering Decoupled Aquaponic Production Systems	D. Allen Davis Interactions of Feed Management and Nutrition in Shrimp Systems	Marlon Greensword Assessing the cost of tilapia production in an airlifted PolyGeyser RAS facility	Jon Buchanan FDA has approved GE salmon – Now what?
10:15 - 10:45	Tetzusan Benny Ron Aquaponics in Hawaii	Jeffrey Terhune Growth trial of Nile Tilapia Fed Combinations of the Probiotic Bio-Cat and Pre-bioitc GroBiotic	JMR Gorle Full-scale CFD modeling with experimental validation of large RAS tank hydrodynamics	Jonas Jonasson Commercially available all-female Atlantic salmon
10:45 - 11:00	Break	Break	Break	Break
11:00 - 11:30	Adam Cohen On the Edge of Water & Energy	Stephano Chu Evaluation of Tissue Lipid and Fatty Acid Composition in Tilapia, Oreochromis spp., Fed on Improved omega 3 and omega 9 Fatty Acid Diets	Ido Seginer A growth model for gilthead sea bream (Sparus aurata L.)	Yonathan Zohar A Novel Approach to Producing Reproductively Sterile Fish for Biological Containment
11:35 - 12:05	Steve Hughes Aquaculture, Aquaponics, and Tutoring Urban Students with Tilapia	Karen Jensen Consumption of Sea Urchin Egesta Enhances Growth and Protein Efficiency of the Shrimp Litopenaeus vannamei in Integrated Co-Culture	Katherin Steinberg The effect of carbon dioxide on metabolism and growth in adult pikeperch (Sander lucioperca)	Jon Buchanan Strategies for the Genetic Improvement of Farmed Fish
12:10 - 12:40	Brian Neerie and Chris Mullins Aquaponic Production in Food Deserts	Tom Zeigler Better Profit Decisions from Simplified Economic Modeling	Esteban JM Emparanza Around the world in 5 RAS farms	Carol Engle Knowing your market: Finfish consumers
12:30 - 2:00	Lunch	Lunch	Lunch	Lunch
	Health and Disease Moderator: Chris Good	Biofloc and Shrimp Moderator: David Kuhn	Biological Control and Biofiltration Moderator: Daniel Taylor	Aquaculture Innovation Workshop
2:00 - 2:30	Chris Good Assessing the effectiveness of peracetic acid to remediate post-vaccination Saprolegnia spp.-associated mortality in Atlantic salmon Salmo salar parr in recirculation aquaculture systems	Geraldo Kipper Foes Removal of Total Suspended Solids In BFT Litopenaeus vannamei Culture in Lined Ponds	Lars-Flemming Pedersen Microbial water quality - tools and challenges	Andrew Tsui and John Davidson Ike jime: Lessons learned from the Japanese method of humane harvesting to optimize the eating quality of Atlantic salmon

Friday, August 19, 2016 (continued)

	Crystal ABC	Roanoke Ballroom A	Roanoke Ballroom B	Washington
	Health and Disease Moderator: Chris Good	Biofloc and Shrimp Moderator: David Kuhn	Biological Control and Biofiltration Moderator: Daniel Taylor	Aquaculture Innovation Workshop
2:35 - 3:05	David Straus Controlling fungus on channel catfish eggs with peracetic acid	Tzachi Samocha Design, operation, and economics of super-intensive biofloc-dominated systems for the production of the Pacific White Shrimp, Litopenaeus vannamei - the Texas A&M AgriLife experience	Paula Rojas-Tirado Effects of feed loading on microbial water quality in RAS	Cathal Dineen and Musleh Uddin A producer and distributor collaborative approach to ensuring great tasting fish for the marketplace
3:10 - 3:40	Thomas Meinelt Disinfection with peracetic acid (PAA), an alternative against fish pathogens	Luis Poersch Super intensive culture of Litopenaeus vannamei in biofloc system: pathways to increase the production in southern Brazil	Keiko Saito Application of anaerobic digestion and microbial fuel cell (MFC) for advanced aquaculture waste management: Production of renewable bioenergy	Samual Chen Hudson Valley Fish Farm's land based steelhead RAS project
3:40 - 4:00	Break	Break	Break	Break
4:00 - 4:30	Dibo Liu Pulse vs continuous treatment: Which is better for applying peracetic acid in RAS?	Guozhi Luo The performance of microbial flocs produced with aquaculture waste as food for *Artemia*	Dallas Weaver Towards zero discharge: Denitrification and refractory organic cleanup	Kirk Havercroft Sustainable Blue's Land Based Atlantic Salmon RAS Project – Salmon Performance Update and Expansion Plans
4:35 - 5:05	Cornelius Becke Consequences of short- and long-term exposure of rainbow trout (Oncorhynchus mykiss) to increased suspended solid load in recirculating aquaculture systems	David Kuhn Ex-situ Biofloc Technology: Review and New Applications	Steven Hall Comparison of rice hulls to EMT media for biofiltration in mesotrophic recirculating systems growing Fundulus grandis	Garry Ulstrom Kuterra's land based Atlantic salmon RAS project
5:10 - 5:40		Rountable Discussion Biofloc and Shrimp Production		Thue Holm Atlantic Sapphire's land based Atlantic salmon RAS projects in Denmark and USA

Poster Sessions

	Poster presenters will be available for questions in the Roanoke Foyer			
5:00 - 6:00	S Chakravarty Case study of an indigenous recirculatory aquaculture system in a fish hatchery of Naihati, West Bengal, India using tilapia as biocontrol agent	ME Megahed Molecular analyses the microbial community dynamics in three RAS with different salinities for production of European eel	N Foore Tracing the Activation Efficiency of the Biofiltration Bacteria in a Recirculating Aquaculture System Using Stable Isotope Analysis	RG Okayi Growth Performance and Nutrient Utilization of Clarias gariepinus (African catfish) Fingerlings Fed Varying Levels of Moringa oleifera (drumstick) Leaves and Seeds
	RH Khalil Some Bacterial and Fungal Affections Causing Disease Problems in Cultured Seabream Sparus surata	D Drahos Identifying Effective Direct-Fed Probiotics in Aquaculture	NB Mahfouz Molecular Characterization of Aeromonas hydrophilia Strains Isolated from Diseased Marine and Freshwater Fish	Ayyat Mohamed Salah Reduction of dietary aflatoxin contamination in fish farm
6:00 - 8:00 pm	**Dinner (Moderator: Dave Kuhn) - Exhibition Hall**			

Saturday, August 20, 2016

8:00AM - 10:00AM	Registration
8:00AM - 4:00PM	Continuous Refreshment Break
9:00AM - 3:00PM	Trade Show Opens

	Crystal ABC	Roanoke Ballroom A	Roanoke Ballroom B	Washington
	RAS Sludge Management Moderator: Tim Pfeiffer and Ron Malone	Pre- and Probiotics Moderator: David Kuhn	Salmonid Diets/Nutrition for RAS Moderator: John Davidson	Aquaculture Innovation Workshop
8:25 - 8:55				Bendik Fyhn Terjesen CtrlAQUA research to optimize RAS for Atlantic salmon post-smolt production
9:00 - 9:30	Todd Guerdat Particulate effluent capture and utilization from recirculating aquaculture systems: Enabling integrated agricultural production systems	David Kuhn Effects of direct-fed probiotics on production, physiology, and the metabolome of fish	Mari Moren Modern Norwegian salmon diets, resource utilization and fish welfare challenges	Colin Brauner UBC InSEAS research program to determine optimal salinity conditions for Atlantic and Coho salmon growth in RAS
9:35 - 10:05	Ron Malone Pneumatic sludge handling in low profile PolyGeyser floating bead filters	Russell Jerusik Optimizing the hatchery production of Cobia, technology to enhance the transition from live feed to dry feed employing liquid feeds and probiotics to improve larval survival	Gary Burr Growth of Atlantic salmon fed alternative protein diets	Lars O.E. Ebbesson Mental robustness indicators for adaptation potential and welfare assessment in fish aquaculture
10:10 - 10:40	Richard Baptiste Characterization of Biofloc from an Ex-Situ Reactor Attached to a Land-Based Integrated Multi-Trophic Aquaculture System	Delbert Gatlin Application of Prebiotics in Recirculating Aquaculture Systems	Alexander Brinker Diet-mediated floating salmonid faeces - benefits for RAS	Christopher Good and Steven Summerfelt Effects of photoperiod on Atlantic salmon post-smolt in Freshwater closed-containment systems (CtrlAQUA PHOTO) and Atlantic salmon growout on a zero fishmeal, zero FIFO, and non-GAMO diet in Freshwater closed containment systems
10:40 - 11:00	Break	Break	Break	Break
11:00 - 11:30	Uri Yogev Fate of feed: Carbon, nitrogen & phosphorous cycles in near zero exchange RAS	Daniel Taylor Application of Direct-Fed Probiotics to Improve Survival in Shrimp and Reduce Pathogenicity of EMS/AHPND Bacteria	John Davidson Effects of feeding fishmeal-free vs. fishmeal-based diet on post-smolt Atlantic salmon performance and product quality and water quality and waste production rates in recirculation aquaculture systems	Beth Cleveland What size to harvest rainbow trout/steelhead in RAS? Consider growth rate, feed conversion, fillet yield, fatty acid deposition, and production efficiency
11:35 - 12:05	Bill Reck NRCS Planning Considerations for Waste Management Systems Available for Recirculating Aquaculture Producers	Roundtable Discussion Trends and Outlook of Pre- and Probiotics in Finfish and Shellfish Production	Mikel Detz Jensen Recirculation feed for Atlantic salmon (Salmo salar)	Jason Mann Trends in sustainable and practical salmon and trout feeds
12:05 - 1:30	Lunch	Lunch	Lunch	Lunch

Saturday, August 20, 2016 (cont.)

	Crystal ABC	Roanoke Ballroom A	Roanoke Ballroom B	Washington
		Hatchery and Fingerling Production Technologies Moderator: Michael Schwarz	RAS Water Quality Control Moderator: Daniel Taylor	Aquaculture Innovation Workshop
1:30 - 2:00		Michael Schwarz Increasing the "Odds" with Copepods	Bertrand Barrut/Philip Fitzpatrick Water delivery capacity, gas exchange and particulate removal efficiencies of a vacuum airlift - application to water recycling in aquaculture system (RAS)	Jillian Fry Examining resource use for feed crops: Challenges and opportunities for aquaculture
2:05 - 2:35		Nick King Managing the Multi-Process (Marine) Hatchery for Health and Production Stability	Tim Pfeiffer Biofiltration capacity of a moving bed biofilter in sequence with a submerged floating bead bed bioclarifier - application to water recycling in aquaculture systems (RAS)	Ragnar Joensen Salmon transfer: Options for pumping entire tank to grade/harvest
2:40 - 3:10		Abode Wilson Omozoya The effect of some water quality parameters on the growth of Clarias gariepinus fed commercial pelleted feed supplemented with live tilapia fry in concrete tank	Lionel Hayter Removal of fine suspended solids in recirculating aquaculture systems using fixed bed contact filters	Frode Mathisen Mortality removal approaches used in Northern European salmon smolt facilities
3:10 - 3:30	Break	Break	Break	Break
3:30 - 4:00		TBA	Aikaterini Spiliotopoulou Efficient ozone transfer into RAS water and an innovative way to determine delivered ozone in site	Steven Summerfelt and Jagan Gorle, Survey – Use of large circular tanks in Norwegian salmon smolt and post-smolt facilities, plus hydrodynamic modelling of such tanks
4:00 - 4:30				Marius Haegh Challenging the status quo – Opportunities for innovation
4:30 - 5:00				RAS Innovation Panel Discussion
5:00 - 5:30				Eric Patel Wrap-Up

Sunday, August 21, 2016

Field trip (Includes breakfast and lunch)

5:15 a.m. - 6:00 a.m. Grab-and-go breakfast for field trip

6:00 a.m. - 6:00 p.m. Freshwater Institute, Shepherdstown, WV

THANKS TO ALL OF OUR SPONSORS!

U.S. AQUACULTURE SOCIETY
A CHAPTER OF THE WORLD AQUACULTURE SOCIETY

aquaLOGIC
Aquatic Design
Manufacturing & Integration

YSI
a xylem brand

PENTAIR

SERVING THE AQUACULTURE INDUSTRY FOR 40 YEARS
aquaculture magazine

SKRETTING
a Nutreco company

panorama acuícola
MAGAZINE

HATCHERY
INTERNATIONAL

Western
Chemical, Inc.

AQUATIC SOLUTIONS

FishVet Group

VEOLIA

COTA
CENTER FOR ORGANIZATIONAL & TECHNOLOGICAL ADVANCEMENT

ICRA Speaker Biographies

Richard Baptiste

Richard Baptiste is a Research Program Coordinator at Harbor Branch Oceanographic Institute at Florida Atlantic University's Center for Aquaculture and Stock Enhancement. His career spans more than 30 years in the aquaculture industry conducting research and production on several species of marine finfish important to sport fishing and food production. For a large portion of his career he conducted research and played a crucial role in production, and technology transfer for Florida's now multimillion dollar shellfish industry. His more recent areas of research encompass work with culture of marine finfish conducted in low salinity systems for USDA's STARR (Sustainable Tank Recirculating Research) and refining technologies for stock enhancement production for valuable salt water game fish in recirculating systems and development land based of Salt Water Multi Trophic Aquaculture systems.

Bertrand Barrut

Bertrand Barrut is a renowned Aquaculture Engineer having authored and coauthored a significant number of papers and literature related to extraction potential of use of Vacuum Column technology in the area of degassing and gas stripping and particle removal in both seawater and freshwater environments. He is responsible for the advancement and development of the technology in aquaculture, algae, wastewater, and oil and gas sectors for Coldep who are the licensees for the technology developed by Ifremer and Insa in France.

Cornelius Becke

Cornelius Becke studied Biosciences at the University of Münster between 2007 and 2012, degree: Master of Science (MSc). Since 2013 conducting PhD at the Fisheries Research Station Baden-Württemberg, topic: "Direct and indirect effects of suspended solid load in recirculating aquaculture systems on the health and growth performance of salmonids".

Alexander Brinker

Alexander Brinker was born in Vechta/Germany in 1973. Studies of biology at the Philipps-Universität Marburg and at the Albert-Ludwigs-Universität. 2000 to 2003 working as research associate at the Fisheries Research Station of Baden-Württemberg (awarded PhD for Rerum Naturalium for studies on suspended solids in flow-through fish farming). From 2004 employed as senior researcher (expert fields: salmonid nutrition, fish parasites, husbandry conditions, mathematical modeling) at the Fisheries Research Station of Baden-Württemberg. Additionally, since 2007 working as lecturer (applied statistics, parasitology and fish biology) at the University of Konstanz. From 2012 ongoing Head of the Fisheries Research Station. Habilitated in 2016 with the Venia legendi in the disciplines of Fish ecology & Aquaculture (Department of Biology, University of Konstanz)

Gary Burr

Gary Burr did his undergraduate work at Jacksonville University and obtained a Master's degree from East Carolina University. He graduated from Texas A&M University in December 2007 with his Ph.D. He studied the effects of prebiotics, non-digestible feed ingredients that affect the intestinal microbial community, on red drum and hybrid striped bass. Currently he works for the Agriculture Research Service in Franklin, Maine as a Research Physiologist, studying alternative feed ingredients for Atlantic salmon and Arctic charr with Dr. Brian Peterson. He also assists with the selective Atlantic salmon breeding program, studying freezing, fillet carotenoid deposition and fillet fatty acid deposition in different salmon families.

Ryan Chatterson

Ryan Chatterson studied Molecular and Micro Biology at the University of Central Florida and has been growing with aquaponics for 15yrs. He spent 10 years working at Aquatic Eco-Systems (Now Pentair) where he assisted in thousands of aquaponic projects ranging from backyard systems to large commercial design. While there he also managed Green Sky Growers roof top aquaponic greenhouse, built and managed 2 large outdoor demonstration systems and helped to design the companies workshop curriculum in which he taught over 150 students alongside Dr James Rakocy, Dr. Wilson Lennard and a few other team members. In early 2013 Ryan left the company to run his own commercial aquaponics farm, Chatterson Farms and in 2014 started Aquaponic Engineering and Design providing design, engineering and educational services for the commercial aquaponic industry. Ryan is a 3 time speaker at IRCA, and has spoken and taught aquaponics at several other conferences and institutions around the world.

Stephano Chu

Hyun Sik Stephano Chu is a Ph.D. candidate in Department of Food Science and Technology at Virginia Polytechnic Institute and State University. He is also a part of Water-Interface Interdisciplinary program. His focused area of research is in food lipid chemistry, tilapia aquaculture, and water chemistry. His dissertation research is on the Evaluation of Tissue Lipid and Fatty Acid Composition in Tilapia, Oreochromis spp., Fed on Improved omega 3 and omega 9 Fatty Acid Diets. Also He have earned M.S. degree in Department of Food Science and Technology with focus on food chemistry and food microbiology of polyphenols in medicinal mushrooms.

Adam Cohen

Adam Cohen holds dual bachelor's degrees in Marine Biology and Marine Fisheries from Texas A&M University, pursued graduate work at the University of St. Thomas in Houston, and has taught mathematics, life and physical sciences, agriculture and entrepreneurship in both private and public high schools for over 12 years. In 2007, Cohen founded Green Phoenix Farms; a company that is focused on providing individuals with the training, materials and designs necessary to produce high quality foods with minimal effort and resource utilization. His commitment to sustainability, natural food and healthy eating, and a concern for the future well-being of communities and the earth as a whole have guided his current efforts to research, develop and educate about Aquaponics. Cohen is committed to developing a method to grow food and aquatic animals in a simple, symbiotic system that can be scaled for "back-yard" hobby farming or up to large-scale profitable commercial production. As a long-term goal, he envisions an urban agricultural model, which will supply natural fruits, vegetables, herbs and a variety of edible fish and other seafood through a totally self-contained and sustainable growing system. Cohen has been a featured presenter on Aquaponics, Urban Food Production and Food Sustainability at speaking to members of the Aquaponics Association, Vocational Agriculture Teachers Association of Texas, Missouri Organic Association, World Aquaculture Society, and Energy OutWest. He lives in Santa Fe with his wife and children.

John Davidson

John Davidson is a Senior Research Associate at the Conservation Fund's Freshwater Institute (TCFFI), in Shepherdstown, West Virginia, where he has been employed for 18 years. John holds an M.S. in Wildlife and Fisheries Resources from West Virginia University and a B.S. in Biology from Shepherd College, WV. He has authored many peer-reviewed articles on a range of topics including: the effects of accumulating nitrate on rainbow trout cultured in recirculating aquaculture systems (RAS); evaluation of alternative protein diets for salmonids; development of procedures to remediate off-flavor from RAS-produced fish; and the production of market-size Atlantic salmon in land-based RAS. John was recently honored in Intrafish's 40 Under 40 series, as an up-and-coming leader in the seafood industry.

D. Allen Davis

D. Allen Davis (Ph.D., Wildlife and Fisheries Sciences, Texas A&M University, 1990) is an Alumni Professor (May 2012 to present) with the School of Fisheries, Aquaculture and Aquatic Sciences. Dr. Davis has dedicated his research and teaching efforts to improve technologies for the culture of marine (e.g. Florida pompano, red drum, red snapper, California yellowtail, white seabass, and Pacific white shrimp) and freshwater (e.g. catfish and tilapia) species for stock enhancement and aquaculture.

Philip M Fitzpatrick

Philip M Fitzpatrick has an 18 year record in the seafood and aquaculture industry both as group managing director Americas of Marine Harvest and in conservation as the Commercial Director Americas of the Marine Stewardship Council more recently he has been active in the development and advancement of new technologies for the food and aquaculture space. He was voted in 2012 as one the globally top 100 most influential people by Seafood International. He currently serves a senior advisor to Searen/Coldep.

Nathan Foore

Nathan Foore has an interest in the optimization of Recirculating Aquaculture Systems, I have focused my studies on analyzing ways of promoting efficiency in the systems. Foore is a current Masters student at the University of Jyvaskyla and about to graduate. He is looking to continue my education in a well suited for PhD program.

Delbert M. Gatlin

Delbert M. Gatlin III is a Regents Professor in the Department of Wildlife and Fisheries Sciences and member of the Intercollegiate Faculty of Nutrition at Texas A&M University. He had an academic appointment with Texas A&M University since 1987. Dr. Gatlin earned a B.S. in fisheries/aquaculture from Texas A&M University in 1980 and a Ph.D. in nutritional biochemistry from Mississippi State University in 1983. He is also a Certified Fisheries Scientist and member of the American Fisheries Society, World Aquaculture Society and American Society for Nutritional Sciences. Dr. Gatlin's research program encompasses many different aspects of nutrition including determination of requirements for and metabolism of various nutrients, as well as development and evaluation of diet formulations and feedstuffs for various fish species including channel catfish, hybrid striped bass, red drum and tilapia. The targeted goal of his research program is to improve sustainability and production efficiency in aquaculture and enhance the quality of resulting products. Gatlin has been Nutrition Section Editor for the journal Aquaculture since 2009. He also was Vice Chair of the Committee on Nutrient Requirements of Fish and Shrimp of the National Academies from 2009-2010. Gatlin has co-authored over 215 peer reviewed journal articles and 16 book chapters.

Christopher M. Good

Christopher M. Good is the Director of Aquatic Veterinary Research at The Conservation Fund's Freshwater Institute (Shepherdstown, WV). Dr. Good earned his B.Sc. from the University of Guelph in Ontario, Canada, and went on to earn his M.Sc., D.V.M. and Ph.D. from the Ontario Veterinary College, Canada. During his graduate studies, Dr. Good worked at the Ontario Ministry of Natural Resources' Fish Health Laboratory and the Canadian Cooperative Wildlife Health Center's West Nile and Avian Influenza Surveillance Laboratory. Dr. Good began work as the Aquaculture Veterinarian at The Freshwater Institute in January 2007, and is a licensed veterinarian in the state of West Virginia and a member of the World Aquaculture Society and World Aquatic Veterinary Medical Association. His current research focuses on improving the sustainability of the aquaculture industry through enhanced health and welfare of farmed fish, and he is involved in peer-reviewed and industry publications, lectures at conferences and workshops, and frequent interaction with stakeholders.

Jaganmohan Rao Gorle

Jaganmohan Rao Gorle is associated with CtrlAQUA SFI as Research Scientist at the Norwegian research firm, Nofima AS. His research interests include hydrodynamics, computational fluid dynamics and design optimization. He received his PhD in Applied Fluid Mechanics from French institute, ISAE-ENSMA. Prior to this, he worked for several industrial and research projects in India and Europe.

Todd Guerdat

Todd Guerdat is a professor of Agricultural Engineering at the University of New Hampshire with appointments in both the Biological Sciences and Civil and Environmental Engineering departments. Dr. Guerdat earned his PhD in Biological and Agricultural Engineering from North Carolina State University researching water and wastewater treatment in recirculating aquaculture systems at the pilot and large scales. Todd has served as the Director of Research and Development for Aquaculture Systems Technologies in New Orleans, LA and as the Director of Nutrient Management Systems at Cambrian Innovation. His agricultural engineering research program at UNH is centered on scalable nutrient capture and reuse systems, protected agricultural systems, and alternative agricultural energy systems which will provide opportunities for students and farmers alike to apply their knowledge in a practical manner. He is currently working toward optimizing the integration of recirculating aquaculture systems with plant production for improved nutrient utilization while preserving economic sustainability. Dr. Guerdat is currently serving as the president of the Aquacultural Engineering Society.

Steven Hall

Steven George Hall, Ph.D. (Cornell), P.E. (NY, LA), has been on the faculty at LSU and the LSU AgCenter, as Assistant, Associate and Full Professor from 2000-present. His focus areas have included Aquacultural Engineering and Coastal Bioengineering, with interests in energy efficiency and conservation, biological systems instrumentation, automated and autonomous systems, resource management engineering and sustainability. He joined the faculty at North Carolina State University (Raleigh) in the Department of Biological and Agricultural Engineering in 2016 and has been named director of the NCSU Marine Aquaculture Research Center at Marshallberg, with teaching, research and extension appointments. He is past president of the Aquacultural Engineering Society, Fellow of the American Scientific Affiliation, and has received numerous teaching, research and outreach awards. He enjoys playing the piano, ultimate Frisbee, growing a garden and making up silly songs with his kids.

Lionel Hayter

Lionel T. C. Hayter is a recent graduate from the Masters of Science in Chemical Engineering program at the University of New Brunswick, Canada where he also received his Bachelor of Science in Chemical Engineering. He is currently employed with Sorensen Engineering in St. Andrews, New Brunswick, Canada, an engineering company with experience in aquaculture design and improvements.

Steven Hughes

Steven G. Hughes has been an active researcher in the generalized topic area of fish husbandry for over 35 years. He received his Bachelor of Science Degree in Biology from the University of Notre Dame while both his Master of Science and Doctor of Philosophy degrees were earned at

Cornell University with combined emphases in nutrition, physiology, and aquaculture. Since 2004, Dr. Hughes has been the Director of the Aquaculture Research and Education Laboratory at Cheyney University of Pennsylvania and is responsible for the planning, initiation and execution of experiments to address research needs in the areas of urban aquaculture, fish nutrition, aquarium fish culture, impact of water quality on fish growth and physiology, and in the control of feeding behavior. The problems addressed are in the broad areas of general fish husbandry (particularly that related to water recirculating aquaculture systems), nutrient sources and use (which includes requirements, energetics, metabolism, and chemosensory aspects), and the impacts of aqueous mineral levels on growth and physiology. In addition, he has recently helped to develop an aquaponics research and production program on the Cheyney University campus which seeks to promote research on plant growth and the expansion of agriculture in urban areas throughout the Mid-Atlantic region.

Mikkel Detz Jensen

Mikkel Detz Jensen holds a MSc degree in biology from University of Copenhagen. Mikkel has been working in the fish feed industry for more than 10 years and has experience within feed formulation, raw materials, water parameters and farm management – especially within recirculation. Mikkel is one of the key persons behind the Orbit diets for recirculation that BioMar has launched globally over the last years. Since 2012 Mikkel has been managing the Baltic

BioFarm team supporting BioMars customers. In BioFarm we strive to optimize our customers results using BioMar products. We discuss feed, feeding, water parameters, recirculation technology and general farm management with our customers.

Nick King

Nick King has worked in the hatchery production of marine fish for 19 years. First as a hatchery manager for GreatBay Aquaculture – the first commercial marine finfish hatchery in the US - and then as a technical manager for hatchery feed suppliers to the global marketplace. Recently, he has transitioned to the aquatic health field where Nick is currently employed by Fish Vet Group (Portland, Maine). His responsibility is to coordinate the laboratory, veterinarian, and commercial teams in providing services and products to the aquaculture and aquatic research markets in North America. Mr. King holds B.S. and M.S. degrees in biology and is specialized in managing many of the processes within the marine fish hatchery.

David Kuhn

David D. Kuhn, research and extension faculty, joined the Department of Food Science and Technology faculty at Virginia Tech after receiving his doctorate, Ph.D., in Civil and Environmental Engineering at Virginia Tech in 2008. He also holds a B.S. degree in Mathematics (Saint Lawrence University), B.S. degree in Civil Engineering (Clarkson University), and a M.S. in Civil and Environmental Engineering (Clarkson University). In his short time as a faculty member he has helped bring in extramural funds (over 30 projects) to the University to work on various aquaculture projects that serve the needs of the aquaculture industry. To date, he has over 140 publications and abstracts. His research interests include animal husbandry, alternative ingredients for aquaculture feeds, biofloc technologies, probiotics, metabolomics, systems engineering, toxicology, and waste handling/reuse. Overall, he enjoys working with industry to help them become more economically and environmentally sustainable.

Dibo Liu

Dibo Liu is a PhD student of Agricultural Science at the Albrecht Daniel Thaer-Institute of Agricultural and Horticultural Sciences at the Humboldt University in Berlin, Germany and in the Department of Ecophysiology and Aquaculture, at the Leibniz-Institute of Freshwater Ecology and Inland Fisheries. Dibo received a B.Sc. of Biological Science from the Faculty of Biology at China Agricultural University and an M.Sc. of Fishery Science and Aquaculture from the Albrecht Daniel Thaer-Institute of Agricultural and Horticultural Sciences at Humboldt University of Berlin.

Guozhi Luo

Guozhi Luo has worked as an Associate Professor for the Department of Fisheries and Life Science at Shanghai Ocean University, Shanghai since 2005. Luo's education includes a B.S. (Aquatic organisms) 1997, an M.S. (Aquaculture) 2000, Shanghai Fisheries University, and a Ph.D. (Environment Science) 2007, Tongji University.

Nadia Mahfouz

Nadia Bassiony Mohamed Mahfouz is a professor and head of Fish Diseases & Management Department on the Faculty of Veterinary Medicine at Kafer el Sheikh University in Egypt.

Ron Malone

Ron Malone serves as a distinguished emeritus professor in the Department of Civil and Environmental Engineering at Louisiana State University where he has served for over thirty-five years. He taught in the area of water and wastewater treatment and in the area of water quality modeling. He is registered as a Civil Engineer and as an Environmental Engineer in his home state. His research has focused on the design of recirculating systems. He has published over hundred technical papers and made over 200 technical presentations. He is the holder of several patents on floating bead filter designs and is a Co-owner and Chief Technical Officer of Aquaculture Systems Technologies, LLC. An AES members since its formation, Dr. Malone previously served on the AES board of directors and as President (2000).

Mohamed Megahed

Mohamed E. Megahed received a PhD and is currently an Aquaculture Researcher at National Institute of Oceanography and Fisheries (NIOF) in Egypt. With 15 years of experience, Dr. Megahed has worked and studied as a researcher at several organizations including the R&D department of National Prawn Company (Saudi Arabia),WorldFish Center (Egypt), Ghent University (Belgium), and AKVAFORSK Genetics Center As (Norway), and also as an Aquaculture Consultant (Africa, Arabian Gulf and Egypt).

Thomas Meinelt

Thomas Meinelt completed his M.A. thesis in fish diseases and earned his Ph.D. in fish toxicology from Humboldt University of Berlin. He is a senior scientist and Deputy Head of the Department of Ecophysiology and Aquaculture at the Leibniz Institute of Freshwater Ecology and Inland Fisheries in Berlin, Germany. His current research fields are fish toxicity of potash mining effluents and alternative treatments in sustainable aquaculture.

Mari Moren

Mari Moren has been working in the field of fish nutrition since 2000. She has worked with several species such as halibut, cod and salmon. Her main scientific interest has been nutrients affecting development of fish larvae. In 2014, she started in Nofima as director of research for the department of nutrition and feed technology.

Mehmet Naz

Mehmet Naz has been studying as Assistant Professor at Iskenderun Technic University, Marine Sciences and Technology Faculty since 2009. His specialty is the focus on the effects of live foods (rotifers and Artemia) and microdiets used in critical larval stages of marine fish larvae such as seabass Dicentrarchus labrax, gilthead seabream Sparus aurata, meagre Argyrosomus regius and common dentex Dentex dentex. In addition, he studied digestive enzymes and hormones of marine fish larvae and live foods. Currently, he is working on microdiet formulations of meagre Argyrosomus regius larvae.

Robert Gabriel Okayi

Robert Gabriel Okayi was born 31st October, 1962 at Idah, Kogi state, Nigeria. He is married and currently a Professor of (Fisheries biology and Aquatic environmental pollution) at the Federal University of Agriculture, Makurdi, Benue state, Nigeria.

Abode Wilson Omozoya

Abode Wilson Omozoya hails from Edo state Nigeria in west Africa. He is a graduate of Zoology (Fisheries) from the University of Maiduguri and is currently running a Fisheries program (M.Sc) from the University of Uyo, Nigeria. He is currently working with Public complaints commission (Investigation officer) and integrated aid initiative as a field officer (Adhoc).

Lars-Flemming Pedersen

Lars-Flemming Pedersen is senior research scientist at DTU Aqua, Denmark. His research areas include water quality and water treatment in recirculating aquaculture systems, application of chemotherapeutic agents, biofilter performance and management practice.

Luis Poersch

Luis Poersch received his bachelor's at Oceanography from Federal University of Rio Grande - FURG (Brazil), master's at Ciencias del Mar from Universidad Católica del Norte (Chile) and PhD at Biological Oceanography from Federal University of Rio Grande - FURG (2004). Dr. Poersch is a Professor/researcher of FURG and has experience in Fishery Resources and Aquaculture, focusing on shrimp production, acting on the following subjects: multi-trophic system, shrimp production in ponds and BFT system, effluent treatment, environment and social impacts.

Bill Reck

Bill Reck, P.E. is the USDA Natural Resources Conservation Service, National Environmental Engineer. Bill has over 25 years' experience in wetland restoration and treatment, design and evaluation of conservation practices, and extensive experience in the area of animal waste management structural design and evaluation of existing structures. Currently, Bill has national responsibility for NRCS's 20 environmental engineering conservation practice standards.

Tetsuzan Benny Ron

Tetsuzan Benny Ron, Aquaculture Specialist with 40 years experience in the agriculture and aquaculture fields and over 25 years as an aquaculture scientist. Aquaculture Professor at the School of Marine Sciences, Ruppin Academic Center. Developed the Aquaculture Training On-Line Learning (ATOLL). Founder of the AquacultureHub (501c3) non-profit with a mission to promote aquaculture by supplying free information. Experienced Aquaculture Extention, Education and Training Specialist at the Department of Human Nutrition, Food and Animal Sciences, College of Tropical Agriculture and Human Resources, and Aquaculture Coordinator at the University of Hawai'i. Former Department Head of Fish Physiology and Genetics at the National Center for Mariculture, Israel Oceanographic and Limnological Research Institute.

Keiko Saito

Keiko Saito is Research Assistant Professor at Department of Marine Biotechnology, University of Maryland Baltimore County at Institute of Marine and Environmental Technology, University System of Maryland. Dr. Saito is a trained biological chemist and environmental microbiologist received her Ph.D. from the Osaka City University. She was a graduate research assistant at University of Michigan, and then received postdoctoral training at Johns Hopkins University and Center of Marine Biotechnology, University of Maryland Biotechnology Institute. Dr. Saito conducts marine and environmental research to create technologies designed to foster the sustainable aquaculture development. She has studied on microbial ecology associated with aquaculture including microbial-mediated waste treatment in marine recirculating aquaculture system (RAS) and prevention of microbial off-flavor production in aquaculture products. Her research interests focus on the microbial nitrogen and carbon cycles in the RAS closed ecosystem by investigating the biology of key microorganisms in nitrification, denitrification, anaerobic ammonia-oxidation (anammox), and anaerobic digestion using molecular and biochemical techniques. Dr. Saito is currently most interested in renewable energy recovery by microbial bioconversion through the waste treatment processes.

Tzachi Samocha

Tzachi Samocha is a recently retired Regents Fellow & Professor with Texas A&M AgriLife Research. After obtaining his Ph.D. from Tel Aviv University, Israel he joined the National Center for Mariculture in Eilat, Israel where he served as Head of the Shrimp Research Unit from 1980 through 1988. In 1989 he joined the Texas A&M University System and for 27 years he served as Director of R&D at the AgriLife Research Mariculture Lab, at Flour Bluff, Corpus Christi, Texas. His research centers on developing sustainable shrimp production practices and transferring this technology to commercial producers. He continues to advance this work through his worldwide consulting activities.

Michael Schwarz

Michael Schwarz is an Aquaculture Specialist at Virginia Tech – Virginia Seafood AREC (Hampton, VA); Adjunct Assistant Professor in the Virginia Tech Fisheries and Wildlife Conservation Department; Adjunct Professor in the Fisheries College & Research Institute; Tamil Nadu, South India; and Graduate Advisory Adjunct Assistant Professor at Southern Illinois University. Michael is Past-President of both the World Aquaculture Society and US Chapter; President of Quantum Tides Inc.; serves on Boards of the Organic Aquaculture Institute, Biosphere Global Institute, and the International Initiative for Sustainable and Biosecure Aquafarming; and is General Manager of the Forever Oceans/Virginia Tech International Aquaculture Academy. Michael is also an Honorary Ambassador to Jeju, S. Korea, a New 7 Wonders of Nature and UNESCO Biosphere Reserve and World Natural Heritage and Global Geopark. Michael's passion is the sustainable development/expansion of global aquaculture through research, extension, education, and industrial development; with ongoing projects in the Americas, Asia, Europe, and the Middle East. His international aquaculture activities incorporate close interactions with government, NGO, academic, private, and public sector stakeholders towards socioeconomic and environmental development within geopolitical contexts.

Ido Seginer

Ido Seginer received his PhD in Agricultural Engineering from Cornell University in 1961. He has worked at Technion, Haifa, Israel in Agricultural Engineering since 1963. In 2002, he became a Retired Emeritus Professor. His current area of interest is modelling and optimal control of bio-systems (horticulture and aquaculture).

Aikaterini Spiliotopoulou

Aikaterini Spiliotopoulou is an M.Sc Environmental engineer with a background in Environmental Science and is currently an industrial PhD candidate dividing time between the DTU and OxyGuard International A/S, working on ozonation of recirculating aquaculture systems. Aikaterini's research during the past 5 year at the Technical University of Denmark (DTU) has been focused on formation of disinfection by-products in swimming pools and water treatment of wastewater effluents. Aikaterini was employed at Water ApS as a development engineer, designing UV reactors for water purification.

Kathrin Steinberg

Kathrin Steinberg has a B.Sc. in Marine Technologies (University of Applied Sciences, Bremerhaven) and a M.Sc. in Sustainable Aquaculture (University of Stirling) and is currently doing her PhD at Christian-Albrechts-Universität zu Kiel and the Association for Marine Aquaculture in Buesum, Germany. Her research is focused on the effect of relevant RAS ongrowing parameters on the growth rates, metabolism and health of the later life stages of adult pikeperch. She also worked for the German Development Agency GIZ as a Junior Advisor in the project 'Promotion of Sustainable Fisheries and Aquaculture' and as a Consultant in Kenya and Ethiopia.

Dave Straus

Dave Straus earned a M.S. in Aquaculture and a Ph.D. in Toxicology from Mississippi State University, and then worked as an industry post-doc at the New Zealand Forest Research Institute specializing in Cytochrome P450 enzyme research and focusing on the effects of pulp-mill effluent on aquatic organisms. Dave works for USDA/ARS studying the applied use of copper sulfate for controlling disease. He has been heavily involved in its FDA-approval process. He also collaborates with colleagues in Berlin, Germany and Hirtshals, Denmark on research designed to introduce peracetic acid as a new disinfectant in aquaculture.

Daniel Taylor

Daniel P. Taylor is a Research Associate with Virginia Tech and co-organizer of ICRA. After receiving his M.S. from Lund University in Sweden and working in industry in Europe and the US, he joined Virginia Tech in 2010 to manage aquaculture research on campus and the SWVA Aquaculture Research Center. His research interests include photobiology, biogeochemical-ecological modeling of aquaculture systems, aquafeed innovation, RAS production engineering, microbial ecology, market development, and diversification of seafood consumption. Taylor enjoys collaborating with international partners, working with candidate aquaculture species, pilot scale research, and helping industry develop applied solutions.

Jeff Terhune

Jeff Terhune earned his Ph. D. in Microbiology in 1997 from Clemson University. He completed his Undergraduate work and Masters of Science from Clemson as well in Aquaculture and Fisheries. He worked for 6 years with Mississippi State University at the National Warmwater Aquaculture Center in Extension and conducting Applied Fish Health Research. He is currently an Associate Professor at Auburn University's School of Fisheries, Aquaculture, and Aquatic Sciences, teaching two Fisheries courses and conducting research and outreach in Applied Fish Health and Disease Control.

Paula Rojas Tirado

Paula Andrea Rojas Tirado graduated as an Aquaculture Engineer in Chile. She did her MSc. in aquatic science and technology at DTU where she focused majorly on water quality and treatments. She is working on her PhD at DTU Aqua - Denmark where her research is based on "Microbial water quality within recirculation aquaculture systems".

Dallas Weaver

Dallas Weaver began designing and building closed aquaculture systems starting in 1973. This business produced over 20 to 25 million fish per year for the research and pet markets in highly automated recycle system (both fresh water and marine). As part of the business, Dr. Weaver has been conducting research on water treatment systems for industry and aquaculture and been able to explore a number of different possible approaches. The invention of fine media fluidized bed biofilters for both waste treatment and aquaculture, the application of packed column re-aeration in aquaculture, the use of pure oxygen system with feedback control in aquaculture, the design development and use of automated feeding system, the use of low cost lime based pH feedback control systems are among the many firsts for Dr. Weaver. Side issues like low cost yeast based rotifer mass production (billion+ per day continuous, stable culture system design) along with various waste treatment issues like contaminated ground water (MTBE contamination) were also solved.

Uri Yogev

Uri Yogev completed a BSc in Geology and environmental sciences, and Biology, and holds an MSc and in Hydrology and Water Quality, from Ben Gurion University of the Negev, Israel. In the last 4 years Uri is working on various types of RASs.

Directory of Exhibitors

Adsorptech, Inc.
452 Lincoln Blvd
Middlesex, NJ 08846 USA
Contact: James Flaherty
Phone: +1-908-735-9528

Email: james.flaherty@adsorptech.com

Adsorptech manufactures oxygen production machines ideal for RAS farm requirements. The EcoGen model has the lowest power consumption than any other oxygen source in the world, is designed with special features that increase reliability thereby avoiding fish mortality caused by oxygen supply disruption, has the lowest cost to install and maintain, avoids the cost of auxiliary equipment and can avoid the high cost of a building or separate enclosure.

Ahlstrom Filtration
122 West Butler St
Mt Holly Springs PA 17065
Contact: Denise Russell
Phone: +1 717-486-6416

Email: Denise.Russell@ahlstrom.com

Ahlstrom is a high performance fiber-based materials company, partnering with leading businesses around the world to help them stay ahead. We aim to grow with a product offering for clean and healthy environment. Our materials are used in everyday applications such as filters, medical fabrics, life science and diagnostics, wallcoverings and food packaging. Ahlstrom's unique Disruptor® technology is creating many new application opportunities across the entire spectrum of water filtration. With the filtration efficiency of ultrafiltration membranes and the pressure drop of a 2 micron nonwoven media, Disruptor® can be considered as an alternative for membranes in many commercial applications. With the ability to remove membrane biofoulants such as organic acids, bacteria and cellular debris Disruptor® is an effective pre-filter to protect membranes from biofouling or as a standalone filter for submicron filtration applications.

Aqua Logic, Inc
9558 Camino Ruiz
San Diego, CA 92126
Contact: Douglas Russell
Phone: +1 858-292-4773

Email: drusssell@aqualogicinc.com

Aqua Logic is an aquatic design, manufacturing and integration company located in San Diego, California. Aqua Logic has proudly served the aquaculture, seafood, aquarium, and zoo industries for nearly three decades and has made it their mission to provide only the highest quality aquatic heating and chilling equipment MADE IN THE USA. Best known for their corrosion-resistant titanium water chillers and heat exchangers, Aqua Logic also designs and builds a wide array of premium aquatic equipment, including complete pump / filtration packages, UV sterilizers, titanium gas boilers, and turnkey rack tank systems for larval rearing, fish holding, and research. With our 30,000 sqft state of the art manufacturing facility we can meet our customer's requirements, big or small. Contact us to discuss your next project.

Directory of Exhibitors

Aqua Production Systems
111 Elshirl Rd RR#2 New Glasgow, Nova Scotia, B2H 5C5
Contact: Philip Nickerson
Phone: +1 902-746-3855
Email: philipnickerson@gmail.com
We design and build water treatment systems for fish hatcheries and lobster pounds. Specialize in heating and cooling.

Aquacal
2737 24th St. North, St. Petersburg, FL 33713
Contact: Stan Crisp
Phone: +1 910 880-2312

Email: stanc@Aquacal.com

AquaCal has manufactured in the United States since 1981 geothermal and air source heat pumps and chillers that are used in aquaculture and aquaponics applications. AquaCal utilizes their patented ThermoLink titanium heat exchanger to eliminate the need for intermediate heat exchangers between the heat pump and the culture's challenging water conditions.

Aquacare Environment Inc
708 Coho Way Bellingham, WA 98225 USA
Contact: Andy Davison
Phone: +1 360 734-7964

Email: andyd@aquacare.com

Aquacare Environment Inc. and JLH Consulting Inc supplies controlled environment aquaculture technology to the aquaculture industry. Our team specializes in intensive land-based controlled environment aquaculture systems. Aquacare and JLH can develop, design and supply complete fish farms for clients worldwide, and can supply cost effective prefabricated system components to existing farms. Pre-engineered and pre-fabricated modular components make farm construction easy and cost-effective. Post construction support services available.

Aquaculture Magazine
203 St. Mary St. Suite 100, San Antonio, TX 78205
Contact: Chris Criollos
Phone: +1 210-903-4743 / 210-504-3642 / 01152 3314660392

Email: crm@dpinternationalinc.com / info@globaldp.es

Aquaculture Magazine is an American publication serving the national and Global Aquaculture industry. Established in 1968, Aquaculture Magazine has over 5,000 subscribers from over 100 countries. Our expertise relies in the content of our publication. We are Knowledge, We are Science, We are Research. We publish innovative techniques and offer insight about current diseases. Reach your unique highly-targeted audience within the aquaculture industry. In short our publication is uniquely positioned to offer an audience that is receptive to your advertising message.

Directory of Exhibitors

Aquaculture System Technologies
108 Industrial Ave New Orleans, LA 70121
Contact: Ron Malone
Phone: +1 504 837-5575

Email: info@astfilters.com

AST is the premier manufacturer of floating bead filters. Our filters provide biological and mechanical filtration using polyethylene media that never needs replacement and is resistant to channeling. Designed for efficiency, AST filters conserve water and use smaller pumps, resulting in energy savings. AST also offers an advanced Polygeyser® line which is auto-backwashing; purging sludge without manual intervention. These filters are designed for long lasting, hands-off operation without using expensive controllers. AST offers filters, aquatic systems, and filtration design services.

Aquaneering Inc.
7960 Stromesa Court
San Diego, CA 92126
Contact: Caitlin Coronel
Phone: +1 858-578-2028

Email: caitlinc@aquaneerign.com

Our Mission . . . We at Aquaneering are passionate about supporting medical research, saving fish species, and supporting aquaculture. We are committed to providing robust housing and filtration solutions for research, conservation, farming, and recreation industries. We strive to continually improve our designs to ensure the health and wellbeing of aquatic species so that our customers can focus on their projects.

Superior Water Quality . . . Because Our Customers Depend On It

aquaticLIFE™ Coatings (a division of Eco Finish®, LLC)
415 Constance Drive
Warminster, PA 18974
Contact: Michael Monti
Phone: 800-333-1443 x752

Email: mike.monti@ecopoolfinish.com

aquaticLIFE™ Coatings virtually eliminate all of the problems of traditional cementitious, fiberglass, and painted surfaces. They can be applied to new or remodel aquatic surfaces to provide a beautiful, long-lasting finish that will not mottle, chalk, peel, or blister. They are available in a variety of different colors, and all are chemical, stain, and fade resistant. Your facilities will love the look and feel, the easy care, and the lower chemical usage.

Directory of Exhibitors

Aquatic Solutions
2963-100th Street Suite 8, Des Moines, Iowa 50322
Contact: Rich Richmond
Phone: +1 515-276-2782

Email: rich@myaquaticsolutions.com

Aquatic Solutions, LLC is an Aquaculture equipment and supply company. We specialize in filtration, aeration, RAS, Fiberglass tanks, Heating, Cooling and water chemistry. We offer Integrity, Innovation, Knowledge, Synergistic Value and Accountability.

Cargill US Aqua
1012 Pearl Street, Franklinton, LA 70438
Contact: Amanda Rosequist
Phone: +1 763-218-8174

Email: amanda_rosequist@cargil.com

Cargill US Aqua is dedicated to aquaculture nutrition. We grow our business through superior service built on over 140 years of experience in understanding and meeting customer needs.

DC International
75, de Vaudreuil
Boucherville, Quebec, Canada J4B 1K7
Contact: Bernard McNamara
Phone: +1 450.449.6400
Email: mcnamara@dc-inter.com

www.dc-inter.com

DC International designs and manufactures forming systems for the construction of large concrete tanks for aquaculture and biomethanisation purposes. Over the last years, DC International has built amongst others tanks in southern Chile for MultiExports Food, in the U.K. for Westport Fish Farm, in Canada for McKenzie Aquaculture and Kuterra's closed containment salmon farm. DC International's technology is fast, economical and can be used by a non-skilled labour. The PVC forming system gives a smooth interior finish with integrated LED lights, if required. The system can be used for circular, octagonal or rectangular tanks up to 16 feet high (5 meters) and there is no limit for the diameter. If used outdoors, insulation can be inserted into the walls before pouring concrete to produce a constant interior temperature. Waterproofing is insured by a hydrophilic patented joint integrated in factory between the PVC profiles that expands once in contact with water.

Directory of Exhibitors

Fish Vet Group
350 Commercial Street
Portland, ME 04101 USA
Contact: Nick King
Phone: +1 207 699 5901
Email: inquiries.na@fishvetgroup.com

www.fishvetgroup.us

Fish Vet Group is the largest provider of evidence-based veterinary services, dedicated diagnostic technologies and innovative aquatic health products to aquaculture producers around the world. From our headquarters in Portland, Maine we drive improvements in the U.S aquaculture industry, offering products and services that ensure environmental health, reduce disease challenges and facilitate trade in live aquatic organisms.

Hayward Flow Control
One Hayward Industrial Drive, Clemmons, NC 27012
Phone: 888.429.4635

haywardflowcontrol.com

Hayward Flow Control, a division of Hayward Industries, is a leading manufacturer of industrial thermoplastic valves, actuation and controls, instrumentation, filters, strainers, corrosion resistant pumps and bulkhead fittings and tank accessories for use in water and wastewater treatment, chemical processing & transfer, chemical feed, aquatic/animal life support systems and general processing systems. Hayward's thermoplastic flow control products can accommodate aggressive and corrosive environments, delicate ecosystems or the strictest chemical balances to keep water and life working together.

Integrated Aqua Systems, Inc.
2867 Progress Pl Ste E
Escondido, Ca 92029
Contact: Sam Courtland
Phone: +1 760-745- 2201

Email: sales@integrated-aqua.com

Integrated Aqua Systems, Inc. is a leading aquaculture systems integrator and equipment supplier. Our product line is comprehensive for recirculating aquaculture including light commercial to industrial mechanical & biological filtration, disinfection equipment, temperature control, pumping and oxygenation. Authorized US Distributor for top manufacturers featuring CM Aqua drum filters, Trojan-Aquafine UV Sterilizers, IAS BioElements and more. Designing, building and supplying systems to meet your specific needs since 2003.

Directory of Exhibitors

Oceans Design
9550 Camino Ruiz
San Diego, CA 92126
Contact: Michael Paquette
Phone: +1 858-602- 9465

Email: michael@oceans-design.com

Oceans Design is a design and manufacturing firm located in San Diego, California serving the aquaculture, and public aquarium industry. We specialize in recirculating aquaculture systems, sea water intakes, control systems, just to name a few. We have a full line of aquaculture specific tanks and filters from our sister company, Gemini Fiberglass. We have taken hatcheries from meager survival rates to over 45% with challenging species like yellowtail with our robust filtration systems and self cleaning tanks. Call us for help with your next project.

Pentair Aquatic Eco-Systems, Inc.
2395 Apopka Blvd.
Apopka, FL 32703
Greg Trusso
Phone: +1 407.461.2384
Email: Gregory.Trusso@Pentair.com

Pentair Aquatic Eco-Systems, Inc. is the largest source of aquatic products and systems worldwide. PAES offers solutions and expertise to improve growing conditions in any environment from recirculating aquaculture systems to improving water conditions in pens. As pioneers in the industry, we provide the best possible solutions for aquaculture facilities, from hatcheries to grow-out and everything in between.

Directory of Exhibitors

PoolPak International
3491 Industrial Drive, York PA, 17406
Tom Scozzari, Vice President of Sales
Phone: 800-959-7725

Email: sales@poolpak.com

With more than 45 years of experience in indoor pool dehumidification equipment manufacturing, PoolPak® International is the most well-known brand in the industry. Our people and products work daily to improve the quality and comfort of indoor pool environments. PoolPak® dehumidification solutions include a variety of heating, ventilation, and air conditioning systems, in addition to an industry-leading PoolPak® support network. For more information, please visit www.PoolPak.com.

Reed Mariculture, Inc.
900 E Hamilton Ave, Suite #100, Campbell, CA 95008
Tim Reed, Dr. Eric Henry
Phone: +1 408-377-1065

Email: TechSupport@ReedMariculture.com

Instant Algae® Marine Microalgae Concentrates for larval fish, bivalves, crustaceans and other filter feeders - Algae When You Need ItTM. Instant ZooplanktonTM pathogen and ciliate free rotifers - Brachionus plicatilis and rotundiformis and Proales similis, a SSS-type 'Naked' rotifer - and copepods Parvocalanus crassirostris and Apocyclops panamensis. Distributor for Otohime TM weaning feeds and ClorAm-X® ammonia and chlorine control.

Skretting
15 Saunders way
Contact: George Demos/ Chris Shilale
Phone: +1 207 591-7077

Email: Chris.shilale@bio-oregon.com

Skretting is the world leader in the manufacture and supply of aquaculture feeds, making it an essential link in the feed-to-food chain. We apply our knowledge of ingredients and the nutritional needs of fish and shrimp to develop innovations that achieve optimum nutritional value, sustainable production and economic performance. We are a pioneer in recirculation feeds and to fulfil our company-wide mission of 'Feeding the Future'.

Directory of Exhibitors

Veolia Water Technologies
250 Airside Drive
Moon Township PA 15108
Contact: TJ Willetts
Phone: +1-412-809-6641
Email: thomas.willetts@veolia.com

Veolia Water Technologies offers a full range of water and wastewater treatment for industry. Veolia's comprehensive technology portfolio includes leading solutions for environmentally responsible and sustainable aquaculture, such as the new RAS2020 Modular Aquaculture System and the Kaldnes® Recirculating Aquaculture System.

YSI Inc.
Address: 1725 Brannum Lane, Yellow Springs, OH 45387 USA
Contact: Darrin Honious
Phone: +1 (937) 767-7241
Email: environmental@ysi.com

www.YSI.com

Zeigler Bros., Inc.
400 Gardners Station Road
Gardners, PA 17324
Contact name: Tom Welch
Phone: +1 717-677-6181

Email: susan.thompson@zeiglerfeed.com

Zeigler Bros., Inc. is a manufacturer of quality animal feed for the aquaculture industry, specializing in finfish, shrimp hatchery and shrimp growout feeds. Capabilities also include custom manufacturing and Zeigler's Technology Transfer Program. Zeigler's quality program is registered to ISO 9001:2008.

The United States Aquaculture Society (USAS) is a chapter of the World Aquaculture Society (WAS). The mission of the USAS is to "provide a national forum for the exchange of timely information among aquaculture researchers, students and industry members in the United States. To accomplish this mission, the USAS sponsors and convenes workshops and meetings, fosters educational opportunities and publishs aquaculture-related materials important to U.S. aquaculture development."

U.S. AQUACULTURE SOCIETY

A CHAPTER OF THE WORLD AQUACULTURE SOCIETY

To join USAS it only costs $5 more than membership in the World Aquaculture Society. Visit the website at http://usaquaculture.org and join the rest of the professionals, academics, suppliers and students who are members. And come to Aquaculture America 2017 in San Antonio TX February 19-22, 2017. Info on the website.

International Conference on
Recirculating Aquaculture
Blacksburg, VA 24061
Phone: 540-231-8643
Fax: 540-231-9293
www.recircaqua.com

Interactions of Feed Management And Nutrition In Shrimp Systems

D. Allen Davis

School of Fisheries, Aquaculture and Aquatic Sciences
Auburn University, Auburn Al 36849-5419
davisd@auburn.edu

ABSTRACT

A significant portion of the shrimp industry has moved to intensive nursery systems and is moving to very intensive productions systems either as traditional clear water RAS or as variants of heterotrophic systems. The feed is the initial source of all pollutants and the driving force for growth, feed conversion and subsequent economic efficiencies. However, the interaction of feeding habits, feed inputs, nutrient density and nutrient loading are often overlooked and poorly understood. We have run numerous studies over the years to demonstrate the interactions of nutrient density, nutrient loading and performance of shrimp. A number of examples will be presented to provide a better understanding of the interaction of feed, feed management and nutrient density on system performance.

When dealing with shrimp one of the first considerations is there feeding habits. Shrimp externally masticate food, which mean feeds are best if they do not crumble. They have evolved as grazing feeders so they are designed to take in relatively small quantities of feed and eat numerous meals in a day. In fact, feed intake is directly related to the length of time food is available, so the longer feed is available or more specifically the higher the number of meal the higher the feed intake. It also appears that when feed is restricted, feed intake improves presumably by forcing them to be neater eaters and pushing them to consume natural foods that may be present in the system. Just as in other species, shrimp are tolerant to a wide range of nutrient intakes or nutrient density of the diet. Hence, they are amenable to a range of protein levels. We have obtained good results with feed with protein levels as low as 16% and as high as 48%. However, the quantity of nutrient entering the culture system are much higher when low density feeds are utilized which means higher demands on nutrient processing. Quite often farmers when faced with poor water quality will reduce the nutrient density of the feed (e.g. protein) to help solve water quality problem (e.g. high TAN levels). This often helps as protein input and consequently N is reduced. However, this is actually counterproductive as it makes for a higher FCR and higher organic loading of the culture system. It is far more efficient to reduce nutrient inputs by shifting feed inputs which not only reduces protein input and consequently N input but it also reduces organic loading instead of increasing it. Furthermore, restrictions in feed force shrimp to eat more efficiently and waste less food. Thus, producing a more efficient system. Given the impact and cost of poor feed management, it is recommended that producers regularly review and improve feed management strategies.

Growth trial of Nile Tilapia Fed Combinations of the Probiotic Bio-Cat and Pre-bioitc GroBiotic

E. Allison[1], K. Naggar[2], D. Davis[1], and J. Terhune[1]

1. School of Fisheries, Aquaculture, and Aquatic Sciences, Auburn University, AL 36830
2. College of Veterinary Medicine , Nursing and Allied Health. Tuskegee University, Al 36088

jsterhune@auburn.edu

ABSTRACT

Two trials were initiated to compare the effects of a probiotic, pre-biotic and a combination probiotic/prebiotic feed in Nile tilapia (*Oreochromis niloticus*). The trials were conducted at the E. W. Shell Research Station of the School of Fisheries, Aquaculture, and Aquatic Sciences in Auburn, Al. In the first trial , 4 experimental diets were used : control (basal diet) , second diet (basal diet supplemented with prebiotic source) , the third diet (basal diet supplemented with probiotic source) and the fourth diet which was the basal diet supplemented with a combination of prebiotic and probiotic sources. The fish (Initial mean weight 5.14 ± 0.10g (p-value =) were divided randomly into 4 treatments for 11 replicates of 20 fish per tank. Fish were feed twice daily with daily feed rations calculated based on % body weight. The growth trial was conducted in a recirculating system with 44 culture tanks (75L Aquaria), water pump, and supplemental aeration (with central line, regenerative blower and air diffusers) in addition to mechanical and biological filtration. During the 10 weeks feeding trial, the fish were fed the experimental diet, body weight, feed intake and survival were measured to evaluate the growth performance. At the end of the experimental period, the fish were counted and weighted, fish samples were taken for proximate analysis. Results (Table 1) showed that there were no notable effects of the dietary treatments on weight gain or survival.

In a parallel study, the evaluation on the effects of the chosen commercial probiotic, prebiotic, and combination of the two was performed in twenty 200 L tanks. Water supply for the tanks was pumped from an adjoining commercial scale heterotrophic culture unit (102,000 L tank with a standing crop of approximately 11,000 Kg of fish) with a return system in place. The experimental tanks were flushed for 1.5 hr, twice per day with the average flow rate into each tank at 6 L/min. Each tank was stocked with 40 Nile tilapia at an average weight of 115 g per fish. This produced a control feed, prebiotic feed, probiotic feed, and a prebiotic+probiotic feed. Tanks (N=20; 5 tanks per treatment group) were randomly assigned one of the four diet treaments described above in Trial 1. The fish were fed to approximate satiation twice per day. Water temperature and oxygen levels were monitored twice per day and total ammonia, nitrite, total alkalinity, total hardness, and pH recorded at least once per week. All fish were removed and weighed every two weeks through week 8 and final harvest occurring at week 12. Mortalities were recorded and were necropsied if suitable for cause of death. Similar to the aquaria trial above, no significant differences in initial weight, weight gain, FCR, or survival were observed between any of the treatments.

Table 1. Aquaria growth trial of Nile tilapia fed either a basal control feed (D1), pre-biotic (D2), probiotic (D3), or both pre-biotic and probiotic (D4).

Diet	Final Biomass (g)	Final Wt Gain (g)	Wt Gain %	Survival %	Feed Intake (g)	FCR
D1	516.77	414.39	404.89	90.45	704.79	1.78
D2	510.26	407.74	397.90	92.73	688.68	1.70
D3	525.29	422.85	413.38	92.27	712.26	1.72
D4	516.02	412.33	398.12	94.09	683.14	1.69
PSE	22.308	22.405	22.191	4.142	16.912	0.074
P value	0.972	0.971	0.955	0.941	0.590	0.780

Better Profit Decisions from Simplified Economic Modeling

Thomas R. Zeigler, Tim Markey and Craig L. Browdy

Zeigler Bros Inc.
400 Gardners Station Rd.
Gardners, PA 17324 USA
tom.zeigler@zeiglerfeed.com

ABSTRACT

Decisions have consequences. Continuous improvement of the processes and procedures by which decisions are made is a desirable objective. This is especially true in the area of profits.

Aquaculture production systems deal with many costly inputs. Capital costs at startup can be quite significant. Facility design and construction and the acquisition of the necessary equipment for efficient and successful operations require up front investments which should not be undercapitalized. As RAS systems increase in popularity there are many options to be carefully considered for automated feeding and for maintaining water quality. These can be quite costly and economic evaluations need to be made so that the final decisions have a high probability of improving profitability.

Feed and seed are two of the most significant components of variable production costs. Both of these inputs can directly and indirectly determine the ultimate success of the crop. Both represent some of the most important investments producers make on an ongoing basis. To maximize profits, good managers must effectively manage costs to assure efficiencies while making prudent investments necessary to achieve shrimp performance levels that drive profits up.

A simple one-page Excel economic model has been developed which allows economic comparisons to be made which can lead to better decisions relating to investments and profitability. The techniques and assumptions made in the use of this model will be discussed, described and demonstrated. This useful tool has proven that the metric, "feed cost per pound of gain," which is frequently used to decide which feed to use, may be misleading and inappropriate for the desired stated purpose. In addition the model can be easily adapted for shrimp or fish production systems.

On the Edge of Water and Energy

Adam Cohen

ABSTRACT

Many issues are facing humankind: providing adequate amounts of food to a growing population, the development of CO_2-neutral, sustainable sources of energy, and the management of water resources represent three key challenges. The concept of the Food-Energy-Water (FEW) nexus will be discussed, and specifically the production of foods (animal and vegetable), fuels and other products for improved human nutrition (microalgae) using methods that are designed to beneficially utilize wastewater, maximize the economic viability of biofuel production, and ensure maximum nutrient recycle for sustainable food production.

Aquaponic Production in Food Deserts

Brian Nerrie and Chris Mullins

ABSTRACT

Fresh affordable quality food items are in short supply in many residential areas resulting in the development of food deserts. Along with food insecurity, food deserts are a major issue facing many states. In Virginia, the Virginia Food Bank Federation reports a 11.8% food insecurity rate or approximately 900,000 people don't know where their next meal is coming from. A recent report from Virginia Tech and Virginia State University pointed out that many Virginia communities have less access to affordable, nutritious food than the national average. One way Virginia Cooperative Extension has focused on addressing this situation is the investigation of small-scale aquaponics in urban environments. Various production systems and crops have been investigated, including decoupled aquaponic production areas. Preliminary data will be presented along with information on production systems that have been tested for Oreochromis niloticus or hybrid grown to 330 g, which supplies 100 g of edible protein per fish, a recommended serving size. In addition to rapid growth, local production of Oreochromis niloticus has many positives including supplying the community with a low fat, low sodium fish of a known source. Another factor in the success of the system was the comparative low price to consumers. Test marketing of whole fish on ice were provided allowing processing to be completed by the purchaser with the cost savings. An educational component of the system includes development and distribution of proper fish handling brochures, food preparation and cold storage. Strict adherence to state and local regulations and required permits is necessary.

Critical Design Criteria for Engineering Decoupled Aquaponic Production Systems

Ryan Chatterson

ABSTRACT

In a world of diminishing resources and increasing populations, recirculating aquaculture affords us the opportunity to sustainably grow high densities of fish in relatively small areas but during the process, significant amounts of waste are created. Aquaponics furthers the sustainability of an aquaculture operation by engineering the waste effluent to grow plants and produce an alternative revenue stream.

To date, the majority of research in the field of Aquaponics has been performed on recirculating or "coupled" Aquaponics systems where the effluent water is delivered to the hydroponic crop and then recirculated to the fish culture tanks in the RAS. While this works well for species that fall within the same water requirements as the hydroponic crop being grown , it can be problematic for species that require higher water quality standards or temperatures outside of those required for plant production.

By properly engineering and decoupling an aquaponic design, the commercial producer gains the ability to independently control both the aquaculture and hydroponic operations allowing the existing aquaculturist the opportunity to focus his expertise while an experienced horticulturist manages the plant operation using standard hydroponic greenhouse management techniques that are already well established within the industry.

Aquaculture, Aquaponics, and Tutoring Urban Students with Tilapia

Steven G. Hughes

Aquaculture Research and Education Laboratory (AREL), Cheyney University of Pennsylvania, Cheyney, PA

ABSTRACT

Aquaculture and aquaponics are extremely useful tools in the classroom as they lend themselves very easily to "teaching across the curriculum." This is particularly true in urban school systems where the novelty of these practices not only serves to educate, but works as a positive attraction to provide excitement and retain interest in the subject matter. Aquaponics and aquaculture integrate seven essential features to the urban classroom: students, professional development for the teachers, technology, living organisms, interaction of multiple curricula, practical application of learned principles, and re-establishing a connection to agriculture. The presentation will include specific examples of how these seven items are important and how these two sciences can be used to involve students and, hopefully, increase the understanding and retention of basic educational principles.

Aquaponics in Hawaii

Tetsuzan Benny Ron

ABSTRACT

Land development for housing and commercial uses is expanding in Hawaii at the expense of agricultural lands. Today, over 90% of Hawaii's food supply is imported. For centuries, Hawaiians developed sustainable means for food production including native taro fields (lo'i) and aquaculture fishponds (Hawaiian: loko i'a).

Modern aquaponics is an additional viable resource to sustainability that combines aquaculture (raising aquatic animals and plants in controlled environments) and hydroponics (growing plants without the use of soil and separated from the ground). This technique relies on fish waste to provide organic food and nutrients which assist in raising vegetables and fruits. The water carries the fish waste that contains toxic nitrogen by-products, such as ammonia, broken into non-toxic by-product by the bacteria, to the plants. The plants' roots absorb the non-toxic ammonia by-products. This process cleans the water that returns to the fish tank. Thus, Aquaponics is a unique Recirculating Aquaculture System (RAS).

Aquaponic gardening has become very popular in recent years as a way to supply fish, vegetables and fruits for domestic use, and numerous aquaponic farms have grown to commercial operation size in Hawaii. By the year 2013, at least five commercial-scale aquaponics ranches were started in Hawaii, growing a variety of vegetables and fish. The most viable products were lettuce and tilapia, which favor the mild tropical weather in Hawaii and therefore are traded in the marketplace year round.

Based on aquaculture and aquaponics we have developed an educational program (ATOLL: Aquaculture Training and Online Learning) that promotes practical skills and knowledge in the practice of aquaculture. The ATOLL curriculum utilizes a variety of faculty expert instructors and has received great praise from current and former students. The ATOLL program promotes sustainability and resource management, and is used in high school (STEM) and university education across the Pacific Islands and the United States.

Part of our training program for aquaponic systems includes training students about the need to pump water from the fish tank to farm or gardening areas. With thoughtful design, appropriate maintenance, and effective management, airlift pumps may allow the transfer of water through an aquaculture and aquaponics system in a more efficient way than using traditional water pumps. Understanding the disadvantages and learning how easy it is to construct an airlift pump can help many aquaponic farmers to reduce their energy cost, while avoiding the high risk and maintenance that results from utilizing a submerged electrical pump in the fish tank.

This presentation will provide specific instructions on the utilization and construction of air-lift pumps for aquaponic farming and introduce participants to our online educational program for use in the community, school, and university training.

Towards Zero Discharge: Denitrification and Refractory Organic Cleanup

Dallas Weaver*[1], Patrick Keane[2], Mark Francis[3], Jimmy Williamson[3], Conal True[4]

[1]Scientific Hatcheries, 8152 Evelyn Cr., Huntington Beach, CA
*Phone: 714-960-4171 Email: deweaver@mac.com
[2] Hatchery Manager, Pacifico Aquaculture, Ensenada, Mexico
[3]Aquaneering, Inc., San Diego, Ca
[4]UABC Marine Studies, Ensenada, Mexico

ABSTRACT

Aquaculture is evolving towards higher levels of water recycle and water efficiency. Cumulative feed burden (CFB), measured as the amount of feed into the system divided by the liquid discharges (including wet sludge) from the system (dry wt feed)/M^3 of discharge is a common way of viewing intensiveness of RAS operation. In recycle aquaculture system technology (RAS) with suspended solids removal (SS), O_2, NH_3, pH, CO_2, temperature and salinity control; nitrate is the final product and it´s concentration tends to build up as the CFB increases along with some refractory organic compounds. By adding denitrification and reducing the NO_3 to N_2 gas we could expect that CFB can be increased to a factor of 10+ thus decreasing water usage. At this point of water consumption, an aquaculture facility becomes relatively independent of the water supply and can be sited at any location desirable for logistics or other reasons. This added flexibility is very valuable for hatcheries, research facilities, SPF facilities and other aquatic facilities where water supply is limited by cost, regulatory, bureaucratic or other constraints.

We have designed, built, tested and installed a denitrification system geared to service multiple RAS systems at a striped bass hatchery in Ensenada Mexico owned by Pacifico Aquaculture, under a situation where the drought on the West coast of N. America effectively restricted the water supply below the point of being able to operate a conventional RAS. The initial NO_3 (N) design load was for 1.8 kg/day. By principle, bacteria can remove NO_3 from water under anaerobic reducing conditions with no oxygen and low ORP readings (-50 to -150 mv). A second treatment step is required to clean up some of the undesirable anaerobic byproducts and other refractor organics with a fine media fluidized bed biofilter. Instead of using the standard methanol or elemental sulfur as a reducing element for denitrifying bacteria, we used common sugar as a carbon source, thus eliminating flammable and hazardous material handling. The entire system was designed and skid-mounted to minimize installation time and effort. This system design approach allowed for actual testing and activation of the media for the biofilters before shipping to the hatchery site. The system was started up and the bioreactors stabilized at the Aquaneering Inc. facility in San Diego CA; at which the microbiology and feedback control systems were stabilized and tuned before shipping to the site. After startup, testing, and validation the system shipped to the site. Having the biological startup and time delays already completed before shipment allowed almost instant startup. Designs and detailed performance information will be presented on both the anaerobic and aerobic phase operation.

Application of Anaerobic Digestion and Microbial Fuel Cell (MFC) for Advanced Aquaculture Waste Management: Production of Renewable Bioenergy

Keiko Saito*, Yonathan Zohar, and Kevin R. Sowers

Department of Marine Biotechnology, University of Maryland Baltimore County (UMBC), at Institute of Marine and Environmental Technology (IMET), University System of Maryland; 701 East Pratt Street, Baltimore MD 21215, USA

*Correspond to: saito@umbc.edu

ABSTRACT

The most advanced recirculating aquaculture system (RAS) treats not only inorganic but also organic wastes generated by fish farming that has been developed and tested for intensive industrial operation. The fully contained RAS exploits several bioreactors including an aerobic biofilter and several anaerobic reactors to remove ammonia, nitrite, nitrate and solid wastes (i.e., fish feces and uneaten food) from culture water by collecting and treating 100% of wastes produced in the system. Conventional anaerobic digestion based bioconversion, well established and widely used technology for sewage and agriculture waste treatment, has successfully adapted in RAS reducing the discharge down to 0.025-0.05% of system volume. In addition, the microbial processes of anaerobic digestion produce biogas (i.e., methane) as by-product that provides cost effective waste treatment. However, anaerobic digestion of marine aquacultural solids has challenged to complete the conversion of chemical oxygen demand (COD) to biomethane, resulted in the maximum bioconversion efficiency at 50%.

Microbial fuel cells (MFCs) represent a new method of renewable energy recovery that directly converts organic matter to electricity using bacteria. MFC has been considered as an alternative option for municipal and industrial waste treatment where anaerobic digestion has been used. Although it was not tested for aquaculture waste management, some of environmental MFC deployment reported promising results including several advantages of MFC over the anaerobic digestion. MFCs are the capability to treat a wide range of waste types, low substrate concentrations, at temperatures below 20°C, and high nitrate/sulfur containing wastes, where anaerobic digestion generally fails to function. To overcome the incomplete treatment by anaerobic digester in RAS, MFC was examined to treat aquaculture waste in fully contained RAS.

It has been known that bacteria could be used to generate electricity, however, it was believed to require expensive materials to shuttle electrons from bacteria to the electrode and supplemental chemicals for the bacterial growth. However, we accomplished the cost reduction of MFC reactor materials and simplified MFC configuration applicable for aquaculture waste treatment. The granular activated carbon (GAC), an inexpensive and durable material, was selected to use as an anode replacing the expensive and fragile carbon cloth and graphite felt anodes commonly used. As GAC provides high surface area (1000 m2/g at minimum), it could greatly improve bacterial adhesion on the electrode. To simplify MFC configuration, single-chamber MFC was designed by removing the expensive proton exchange membrane (PEM) employed to separate anode and cathode chambers. Single chamber MFCs have been resulted in higher power output compared to traditional two-chamber MFCs. In addition, biocathodes using bacterial catalytic activities of biofilm on the stainless-steel brush electrode was successfully employed instead of expensive platinum cathode. Of two types of biocathodes; 1) Aerobic biocathodes using oxygen as the oxidant assisted by microorganisms; and 2) Anaerobic biocathodes using compounds such as nitrate, sulfate, iron, manganese, selenate, arsenate, urinate, fumarate and carbon dioxide as terminal electron acceptors, an anaerobic biocathode with nitrate as an electron acceptor was tested due to the nitrate removal capability (Equation 1).

$$2NO_3^- + 10e^- + 12H^+ \rightarrow N_2 + 6H_2O \; (+0.74V) \qquad (1)$$

A novel design of MFC, a single-chamber reactor integrating GAC anode and anaerobic denitrifying biocathode was first tested in the pilot scale MFCs (Fig. 1) that successfully generated electricity via treatment of biodegradable wastes produced in RAS without any supplemental materials. The PEM-less MFC with inexpensive electrodes favored for scale-up in any waste treatment has been examined for in-line continuous flow operation by integrated in the fully contained RAS with an up-flow anaerobic sludge blanket (UASB) reactor during intensive fish production. Application of MFC resulted in the increase of COD and nitrate removal rates in RAS overall. Continued study has been focusing on increasing the bioconversion rates of anaerobic digester and MFC in RAS by optimizing the operational parameters for consistent treatment at large scale that will be also reported.

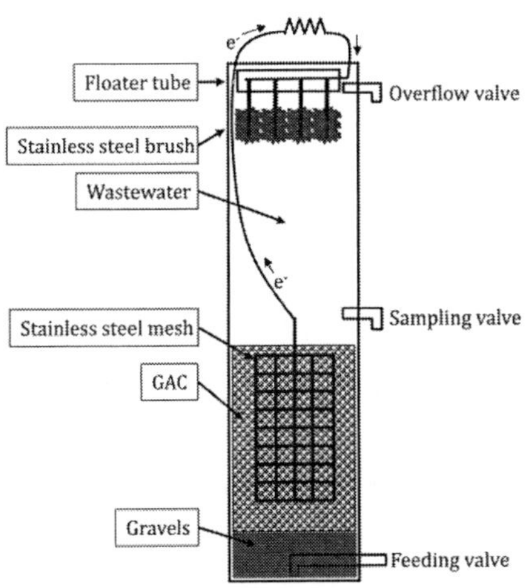

Fig. 1. Configuration of pilot single-chamber microbial fuel cell (MFC) with granular activated carbon (GAC) anode and stainless steel brush aerobic biocathode. Pilot MFC reactor (20 L) contains 6 L (2.4 kg) of GAC. Gravels are used to separate feeding line and packed GAC.

Microbial Water Quality - Tools and Challenges

Lars-Flemming Pedersen* & Per Bovbjerg Pedersen

Technical University of Denmark, DTU Aqua, Section for Aquaculture,
The North Sea Research Centre,
P.O. Box 101, 9850 Hirtshals, Denmark
*Corr. author Email addresses: lfp@aqua.dtu.dk

ABSTRACT

There is a steadily increasing interest for operational tools to assess microbial water quality. Deeper insight into the microbial water quality will increase process understanding and allow potential optimization of RAS management.

Microbial water quality (e.g. diversity and composition of bacteria, bacterial abundance and activity), is under the influence of a sum of factors of which the majority demands on further investigation to predict potential causal relationships. Operational tools to monitor, ideally simple and quick methods, can be used to ensure safe and stable baseline conditions, to identify sudden changes (deviations from baseline) and potentially contribute to improved system performance by identifying suboptimal treatment component or practices.

This presentation includes results from a case study about microbial water quality assessment from a commercial RAS. A 24-hour sampling program was set up to monitor diurnal changes in connection to deteriorated water quality. An array of analysis were performed on the location, including pH, oxygen, TAN, nitrite, nitrate, suspended solids, chemical oxygen demand, turbidity, bacterial activity assessment and particle size determination on fresh samples. The study showed a two-three-fold reduction in microbial related measures over 24 hours as water quality improved following cease of feeding. Strong collations between particulate COD, turbidity, Bactiquant® values and bacterial activity quantified as indirect catalase activity were found. Stable nitrate levels (range: 72.6-76.0 mg nitrate-N/l) throughout the period confirmed that dilution was not the reason for water quality improvement.

In addition to the full scale case study, controlled batch experiments were made to investigate single factor effects on microbial activity. Examples on bacterial activity and regrowth dynamics in RAS water exposed to different concentrations of formaldehyde will be presented and potentials and limitations of the microbial assays applied in the present studies will be discussed.

Effects of Feed Loading on Microbial Water Quality in RAS

Paula A. Rojas-Tirado[1*], Per Bovbjerg Pedersen[1], Olav Vadstein[2], Lars-Flemming Pedersen[1]

[1]Technical University of Denmark, DTU Aqua, Section for Aquaculture, The North Sea Research Centre, P.O. Box 101, DK-9850 Hirtshals, Denmark. *Corr. author Email addresses: part@aqua.dtu.dk

[2]NTNU Norwegian University of Science and Tecnology, NTNU, Department of Biotechnology, N-7491 Trondheim, Norway.

ABSTRACT

Feed composition and feed loading have significant impact on the physico-chemical and microbial water quality for the fish in recirculating aquaculture systems (RAS). Still relatively few studies have been done on microbial water quality, including abundance, activity and composition of microbes in the water, despite the fact that detrimental fish-microbe interactions are a problem in RAS. Intensive RAS water evolves gradually from being oligotrophic to eutrophic and saprobic due to dissolved nutrients, dissolved organic matter and micro particles accumulating within the systems, creating ideal conditions for bacterial growth.

In the present study we evaluated the consequences of changes in feed loading on bacterial abundance and activity under three controlled scenarios in duplicates: i) ceased feeding (0 g feed/d), ii) 250 g feed/d (control) and iii) increased load (500 gram feed/d) during an 11 weeks period. The experiment was conducted in six independent and identical pilot-scale RAS, each having a total volume of 1.7 m^3 and operated with a fixed amount of 80 l/d make-up water. Each system was stocked with approximately 66 kg/m^3 of rainbow trout juveniles (*Oncorhynchus mykiss*). Prior to trial start-up a three month period with constant feeding of 250 g feed/d, ensured systems equilibrium/steady state. Physico-chemical water quality parameters (chemical oxygen demand, biological oxygen demand, total suspended solids, volatile suspended solids, particle size distribution and number, ammonium, nitrite and nitrate concentrations) were measured weekly together with fish performance assessment. Microbial water quality was evaluated by i) bacterial activity via the Bactiquant® method and ii) bacterial abundance quantified by flow cytometry.

Results showed that bacterial activity and abundance was considerably influenced by the changes in feeding and so were physico-chemical water quality parameters and fish performance. The main results and findings of this study will be discussed in the presentation, as well as aspects of microbial water quality in RAS.

Ex-situ Biofloc Technology: Review and New Applications

David Kuhn, Holly Packard, Pedro Ivo Guimaraes, Addison Lawrence, Ann Stevens, Ryan Senger

ABSTRACT

Biofloc technology for aquaculture has been demonstrated to offer numerous benefits for shrimp and species of finfish such as tilapia. There are two primary biofloc technology systems that can be considered for aquaculture production, in-situ and ex-situ biofloc technology. In-situ bioflocs are formed in the culture pond/tank by manipulating the carbon to nitrogen ratio (C:N). Ex-situ biofloc technology is when bioflocs are generated external to aquaculture production and bioflocs are used as a supplemental ingredient in fish or shrimp diets. The focus of this presentation will be on ex-situ biofloc technology.

Ex-situ bioflocs are produced in suspended-growth biological reactors to remove suspended solids and nutrients in organic effluent waters. Carbon supplementation can also be used to promote biological activity to generate a healthy heterotrophic bacteria population. Under this scenario, bioflocs are produced in the biological reactors and the bioflocs can then be used as a feed supplement for fish or shrimp. Ex-situ biofloc technology is a newer technology that is gaining attention by many industries. For example, in our laboratories we have evaluated ex-situ biofloc technologies to treat effluent waters from fish farms, food processing operations, and most recently effluent from an ethanol processing facility. Results from these studies will be discussed.

Removal of Total Suspended Solids In BFT *Litopenaeus vannamei* Culture in Lined Ponds

Geraldo Fóes, Julio Zemor, Edu de Oliveira Holanda, Luis H. Poersch, Wilson Wasielesky Jr.

Federal University of Rio Grande - Institute of Oceanography - Rio Grande - RS, Brazil

ABSTRACT

In intensive shrimp production using the BFT system the water used to be minimal or not exchanged. Therefore, suspended particulate matter tends to accumulate during the production cycle. The TSS can affect water quality and consequently shrimp performance. In order to remove TSS excess is required to use decantation techniques, filtration or water renewals. In this context, the clarification of water during production through the use of clarifiers can be an alternative in the removal of suspended solids, which can generate smaller volumes of effluents and economy of water.

The objective of this study was to evaluate the efficiency of the use of clarifiers in 600 m^3 lined ponds and their effects on water quality and shrimp production performance. The evaluation of the efficiency of clarifiers was carried out in ponds containing algal dominance, and low heterotrophic predominance. The TSS were maintained lower than 500 mg/L using clarifiers with 0.5 and 1.0 % of the pond volume.

Litopenaeus vannamei juveniles (0.83 ± 0.41 g) were stocked (87 shrimp/m^2) in nine growout ponds. The treatments were: (1) Renewal (using water renewals for maintenance of solid); (2) Clarification 0.5 % (clarifier 3,000 L, corresponding to 0.5% of the pond volume) and (3) clarification 1.0 %, two clarifier of 3,000 L interconnected totaling 6,000 L, (1.0 % of the pond volume). The data were statistically analyzed, through Levene's Test to check the variances homogeneity, Anova one-way and Tukey's Test ($p<0.05$) to detect possible differences among treatments.

The experiment lasted 105 days. The water quality values are shown in Table 1. The values of zootechnical parameters can be observed in table 2.

Table 1. Average values (± SD) of water quality parameters in growout of *Litopenaeus vannamei* in lined ponds.

Treatment	T (°C)	DO (mg/L)	pH	NAT (mg/L)	SST (mg/L)	Effluent (L/kg shrimp)
Renewal	25.8 ± 2.3	7.35 ± 0.89	8.63 ± 0.30	0.30 ± 0.29	366 ± 88	1128,8[b] ± 119,0
Clarifier 0.5 %	25.8 ± 2.2	7.48 ± 1.00	8.62 ± 0.32	0.12 ± 0.06	397 ± 56	35,4[a] ± 6,5
Clarifier 1.0 %	25.8 ± 2.3	7.39 ± 0.81	8.48 ± 0.30	0.08 ± 0.12	370 ± 53	50,5[a] ± 18,6

Table 2. Average values (± SD) of the zootechnical parameters of *L. vannamei* after 105 days of experiment.

Treatment	Initial weight (g)	Final weight (g)	FCR	Weekly growth (g)	Productivity (kg/ha)	Survival (%)
Renewal	0.83 ± 0.41	12.74 ± 2.75	1.32 ± 0,09	0.79 ± 0.05	9,212 ± 715	83.7 ± 3.4
Clarifier 0.5 %	0.83 ± 0.41	12.14 ± 2.35	1.32 ± 0,06	0.75 ± 0.07	8,721 ± 482	83.3 ± 4.4
Clarifier 1.0 %	0.83 ± 0.41	11.89 ± 2.34	1.32 ± 0,07	0.72 ± 0.06	8,694 ± 473	86.1 ± 5.7

The water quality parameters did not presented significant differences among treatments ($p>0.05$), and were not limiting for *L. vannamei* growth and survival.
The results showed that removal of suspended solids by clarification using clarifiers was effective for the maintenance of TSS below 500 mg L^{-1}. However, the amount of effluent renewal treatment was superior ($p<0.05$) to treatments with clarification.

It was concluded that the use of clarifiers in systems dominated by algae is feasible, and clarifiers with 0.5% of pond volume is effective to maintain the TSS levels lower than 500 mg/L. And besides, the use of clarifiers generates a smaller amount of waste water and lower water use during the production cycle.

Acknowledgements
The authors are grateful to The Brazilian Council of Research (CNPq), FAPERGS, CAPES, Ministry of Fisheries and Aquaculture (MPA)–Brazil, Centro Oeste Rações S.A. (Guabi), All-Aqua Aeration, Aquatec, Trevisan and Inve do Brasil for funding this research.

The performance of microbial flocs produced with aquaculture waste as food for Artemia

Guozhi Luo, Miaolan Yao, Hongxin Tan, Nan Zhang, Shengkai Wu
Shanghai Ocean University

ABSTRACT

The use of recirculating aquaculture systems (RASs) can recapture the wasted nutrients, which provides an opportunity to reuse them. Recently, biofloc technology (BFT) has been proposed to convert the nutrients in the fish waste into microbial flocs under aerobic conditions. Suspended growth reactors (SGRs) have been employed to produce bioflocs using aquaculture waste. Performance of the flocs, which were produced using solid waste from a recirculating aquaculture system stocking *Anguilla Anguilla*, as food for *Artemia* was investigated over a 18-day feeding trial. Four dietary treatments were set up: *Chlorella* only (diet 1), flocs only (diet 2), and both *Chlorella* and flocs offered as mixed diets in different proportions (diets 3 - 4). The survival rate of diet 1 and 4 were significantly higher than that of diet 2 and 3. Individual length (10.02±2.44 mm) and biomass production of diet 3 (3.27±0.40 g L^{-1}) were highest among the four diets. The crude protein contents (dry *Artemia* $weight^{-1}$) for diets 2, 3, and 4 were 59±3%, 58±2% and 53±3%, respectively, which are significantly higher than that of diet (15±3%). The compositions of free amino acids for the four diets were the same. The fatty acid composition for the four diets was the same except for C24:0. The highly unsaturated (HUFA) in diet 2 was higher than that of the other three diets. The present study showed that flocs produced from fish waste can be used as a nutrient source for *Artemia*. This re-use of otherwise wasted nutrients helps intensive aquaculture to become more sustainable.

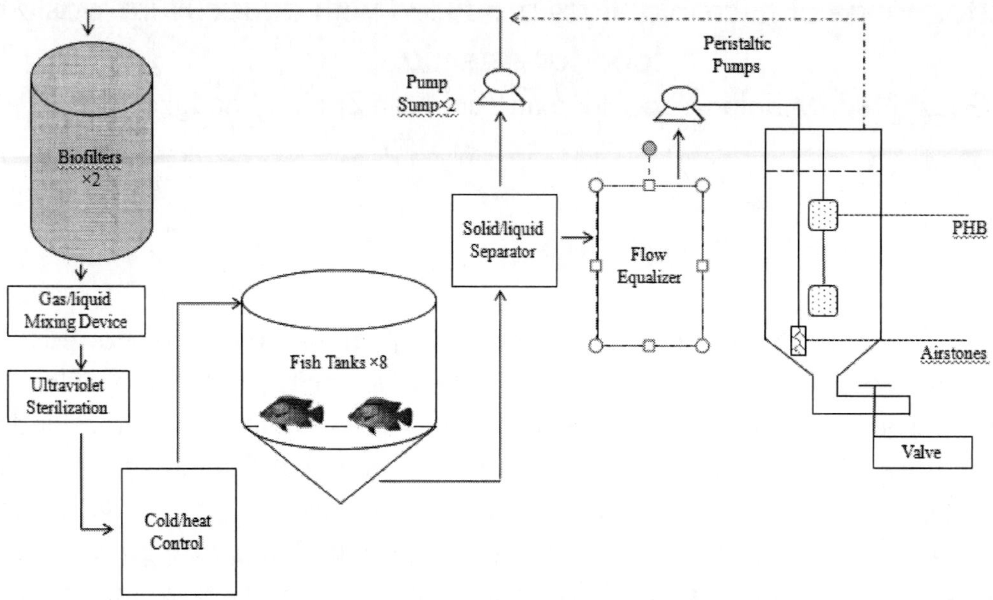

Fig. 1 Simplified experimental set-up, comprising a closed commercial *Anguilla anguilla* cultured system with eight fish tanks in parallel (only one tank is pictured) and nine microbial floc production reactors (only one reactor is pictured) connected to a solid/liquid separator

Increasing the Odds with Copepods

Michael H. Schwarz*[1], Jerome Bosmans[2], Chad Clayton[3], and Stephen Urick[1]

1 Virginia Tech
2 IDEE Aquaculture, www.ideeaquaculture.com
3 Reed Mariculture Inc., www.reedmariculture.com

ABSTRACT

Aquaculture continues to grow in leaps and bounds across production sectors, countries, and continents. As aquaculture production increases, so does the importance of hatchery capacity and capabilities. In the production of many species, there is reliance during the hatchery/larval phase for some form of live feeds. This is ever more important in the marine sector, with emerging high value food fish and ornamentals not only demonstrating a need for traditional live feeds such as rotifers and Artemia, but also the use of copepods in the live feeding regime. There are three groups "phylogenetic orders" of copepods of particular interest in aquaculture; these are: Calanoida, Cyclopoida, and Harpacticoida. Among others, some of the reasons for increasing use of copepods during larviculture (to achieve reliable high survival and quality weaned fry) relate to their (early nauplii stage) small size, attractive movement in the water column, high digestibility, and often superior nutritional value. Presented will be some general background information regarding copepods, some pros and cons relating to their use and culture, followed by examples of culture and application.

Managing the Multi-Process (Marine) Hatchery for Stability in Health and Production Performance

Nick King

Fish Vet Group
350 Commercial Street Portland, Maine 04101.
1-207-699-5909.
nick.king@fishvetgroup.com

ABSTRACT

Expansion of marine or freshwater finfish aquaculture will require greater hatchery production of juveniles, and the ability to produce these juveniles in any geographic location with supply being regular and consistent. Fry survival in the marine fish hatchery to juvenile stage rarely exceeds 50%, and this for commercially developed species. Often with new species development, or species having a new experience with the intensive production system, fry survivals are generally less than 20%, and through applied research "on the farm", or via trial and error experience, these survival number will increase over time until full commercialization. The goal of full commercialization is to not only optimize survival, or maximize production per unit volume in an economical way, but to have a consistent and stable result cycle after cycle in a given production season.

Consistency and stability come from a program of quality management in the hatchery. The hatchery needs to be managed according to performance standards, minimizing risk, and creating stability in the program. New protocols must arise from a methodical approach initiating investigation from a predictable, calibrated system. Otherwise, the results are too random, and the protocol risks becoming unnecessarily complex and open to instability.

An evidence-based hatchery health plan can bring to light the landscape of pathogens present in the facility, and the risks they present. Pathogen risk may also impact performance standards and program stability even if they may only be in the background (i.e. parasites) and not risen to the level of disease outbreak. A hatchery health monitoring program surveys the facility for underlying bacterial, viral, parasitic and non-infectious disease at critical points in the production cycle. This provides the foundation for biosecurity decisions that will ultimately minimize risk.

Consequences of short- and long-term exposure of rainbow trout (*Oncorhynchus mykiss*) to increased suspended solid load in recirculating aquaculture systems

Cornelius Becke[1]*, Mark Schumann[1], Dieter Steinhagen[2], Alexander Brinker[1]

[1]Fisheries Research Station, Argenweg 50/1, 88085 Langenargen, Germany
[2]Fish Disease Research Unit, Centre of Infectious Diseases, University of Veterinary Medicine Hannover, Bünteweg 17, 30559 Hannover, Germany
*Corresponding author: Cornelius.Becke@lazbw.bwl.de

ABSTRACT

Particle load is regarded as one of the most difficult problems in recirculating aquaculture systems (RAS). Despite technical advances in the treatment of solid wastes, invariably a fraction of small particles remain in the system and accumulate over time. While little is known about the direct impact of these accumulating particles on stress, performance or health parameters of fish in RAS, high levels of suspended solids, especially fines, are widely considered harmful to stock health and welfare.

A short-term and a long-term study were performed to investigate the impact of accumulating fines on rainbow trout (*Oncorhynchus mykiss*) in RAS. A unique and decisive criterion of these studies was the isolation of particle accumulation effects from the many other confounding or interacting factors that generally accompany increased particle load, such as levels of ammonium, nitrite, pH and CO_2

Rainbow trout were housed in two replicate RAS. The particle load in one system was artificially increased by a multiple factor, while the parallel system remained under normal conditions as a control. The impact of accumulating particles on fish health was examined using a wide range of parameters (heat shock protein 70, plasma cortisol, hematological parameters, fin condition, etc.). Particle accumulation was monitored in terms of TSS concentration, number of particles per size class and particle size distribution. In a novel approach, the shape and structure of particles was also investigated by using a Particle Insight Size and Shape Analyzer and digital image analysis.

In the short-term experiment (4 weeks), suspended solid load increased in the treatment RAS to over 30 mg TSS/L, while the sister RAS exhibited a maximum suspended solid concentration of 5 mg/L. By the end of the study, the concentration of particles smaller than 32 μm amounted to 0.3 ± 0.3 mg/L in the control system and to 8.0 ± 2.7 mg/L in the treatment system. Increased particle load influenced fish performance only in terms of feeding behavior: fish in the treatment group showed a calmer, less aggressive behavior during feeding compared to the control. Otherwise, there was no apparent effect of elevated fines on fish performance. Gill structures did not show any effect of increased suspended solid load and remarkably, by some markers, gill damage appeared less severe in the treatment RAS than in the control. Furthermore, none of the stress markers and hematological parameters assessed suggested any negative consequences for fish in the treatment group. Overall, the study was unable to show any negative impact of a short-term increase in suspended solid load up to 30 mg TSS/L on fish health and performance or on system stability. In the long-term experiment, rainbow trout were exposed to increased suspended solid load until they reached slaughter weight. In addition to the investigation parameters of the short-term study, the effect of increased suspended solid load on bacterial population was also assessed, by determination of total bacteria count and bacterial load of the gills. The findings of this experiment will be presented.

Assessing the effectiveness of peracetic acid to remediate post-vaccination *Saprolegnia* spp.-associated mortality in Atlantic salmon *Salmo salar* parr in recirculation aquaculture systems

C. Good[1], J. Davidson[1], D. Straus[2], W. Wolters[3], B. Peterson[3], L-F. Pedersen[4], V. Phuntumart[5], C. Lepine[1], and S. Summerfelt[1]

[1] The Conservation Fund's Freshwater Institute, 1098 Turner Road, Shepherdstown, WV 25443, USA.
[2] USDA-ARS Stuttgart National Aquaculture Research Center, 2955 Hwy East, Stuttgart, AR 72160, USA
[3] USDA-ARS National Cold Water Marine Aquaculture Center, 25 Salmon Farm Road, Orono, ME 04469, USA
[4] Danish Technical University, National Institute of Aquatic Resources, Nordsøen Forskerpark, 9850 Hirtshals, Denmark
[5] Bowling Green State University, Department of Biological Sciences, Bowling Green, OH 43403, USA

ABSTRACT

Disease is a major barrier to aquaculture production worldwide, and within the salmon industry it is responsible for the majority of market supply fluctuation. Ubiquitous oomycetes of the *Saprolegnia* genus are particularly problematic disease agents, associated with an estimated 10% mortality among all hatched farmed Atlantic salmon. A major risk period for saprolegniasis is during the weeks following vaccination, the associated stress and tissue damage of which favors the opportunistic *Saprolegnia* spp. to establish infection and cause clinical disease. Atlantic salmon smolt production is increasingly being carried out in recirculation aquaculture systems (RAS), and therefore effective therapeutic strategies must be developed, not only to reduce losses to saprolegniasis, but also to maintain adequate water quality for fish health and performance, i.e. to not impact RAS biofiltration capacity. There is a clear need to develop novel strategies for saprolegniasis control that are both efficacious and not detrimental to RAS biofilter performance. As such, we are aiming to assess daily low-dose peracetic acid (PAA) treatment regimens to determine their effectiveness in reducing post-vaccination losses to *Saprolegnia* spp. infections while assessing biofilter performance. Twelve replicated RAS (12 x 0.5m^3 circular tanks), each with an associated small fluidized sand biofilter, will each contain 200 Atlantic salmon parr (~40g). A portion (600) of these fish will be PIT-tagged to identify family group; these fish will be in the control tanks, in an effort to estimate heritability of saprolegniasis resistance. The fish will be acclimated for one week following arrival, at which point they will be anesthetized (75 mg/l MS-222) and vaccinated with a commercially available salmon vaccine (furunculosis and vibriosis) via intracoelomic injection following standard industry protocols. Following vaccination, the twelve RAS will be randomly assigned to four treatment groups (n=3), receiving either daily single-dose treatments with i) 0.15 mg/L PAA; ii) 0.5 mg/L PAA; iii) 1.0 mg/L PAA; or iv) no treatment. During the three-week post-vaccination period, daily mortalities will be quantified (with mortalities from control tanks recorded for PIT tag information) and assessed for saprolegniasis via wet-mount microscopy. Survival curves for each treatment will be developed and compared statistically. Thrice weekly water quality samples will be collected and assessed for TAN, NO_2-N, and NO_3-N to monitor biofilter performance. Twice weekly water samples will be collected and assessed for *Saprolegnia* spp. counts and oomycete genetic sequencing. Pre-study and weekly samples of three fish per RAS will be euthanized (200mg/L MS-222) and gill tissue will be collected for histopathology in order to assess fish welfare under the treatment conditions. Data collection will be completed by mid-May, 2016, and results will be presented at ICRA in August, 2016.

Pulse vs. Continuous treatment: which is better for applying peracetic acid in RAS?

Dibo Liu[a], Lars-Flemming Pedersen[b], Carlo Cabacang Lazado[b], David L. Straus[c], Thomas Meinelt[a]

[a]: Leibniz-Institute of Freshwater Ecology and Inland Fisheries, Mueggelseedamm 301, 12587 Berlin, Germany. Email: liu@igb-berlin.de, meinelt@igb-berlin.de

[b]: DTU-Aqua, National Institute of Aquatic Sciences, Section for Aquaculture, North Sea Research Centre, P.O. Box 101, DK-9850 Hirtshals, Denmark. Email: lfp@aqua.dtu.dk, claz@aqua.dtu.dk

[c]: U.S. Department of Agriculture, Agricultural Research Service, Harry K. Dupree—Stuttgart National Aquaculture Research Center, P.O. Box 1050, Stuttgart, AR 72160, USA. Email: dave.straus@ars.usda.gov

ABSTRACT

Peracetic acid (PAA) is a promising disinfectant in aquaculture. It is highly effective against various fish pathogens and also environmentally friendly due to harmless degradation residues. However, knowledge about potential adverse effects of PAA disinfection on fish is lacking; practical guidelines for applying PAA in recirculating aquaculture systems (RAS) require further research. In the present study, we compared two treatment strategies for applying PAA in tanks containing rainbow trout (*Oncorhynchus mykiss*): pulse batch of 1 mg/L PAA (twice per week) and a continuous drip of 0.2 mg/L PAA. We monitored the stress condition of fishes for 4 weeks by measuring cortisol concentration in the outflow. After 6 weeks of treatment, we sampled fishes and measured antioxidant in gill and blood. The anti-protease activity was measured in blood and skin. Biofilm was reduced by the pulse batch treatment, but enhanced by the continuous drip treatment. Fish were stressed by the first pulse batches and showed adaptation afterwards by lowering the stress response to negligible levels. In contrast, stress was not induced by the continuous drip treatment. The pulse batch treatment resulted in a larger increase of total antioxidant in serum than the continuous drip treatment. However, only the continuous drip treatment stimulated an increase of antioxidant in gill tissue. The anti-protease activity in serum was reduced by the continuous drip treatment as compared to the pulse batch treatment. Moreover, we demonstrated the importance of hydrology on the practical application of PAA with a mathematical model. In conclusion, we recommend that a pulse batch treatment of PAA, with sufficient intervals between treatments, is more manageable and less stressful for the fish and biofilter in RAS than the continuous drip treatment.

Controlling Fungus on Channel Catfish Eggs with Peracetic Acid

David L. Straus[a] and Thomas Meinelt[b]

[a] U.S. Department of Agriculture, Agricultural Research Service, Harry K. Dupree—Stuttgart National Aquaculture Research Center, P.O. Box 1050, Stuttgart, AR 72160, USA. Email: dave.straus@ars.usda.gov
[b] Leibniz-Institute of Freshwater Ecology and Inland Fisheries, Mueggelseedamm 301, 12587 Berlin, Germany.

ABSTRACT

There is much interest in the use of peracetic acid (PAA) to treat pathogens in aquaculture. It is a relatively new compound and is approved for use in Europe, but not in the United States. This study determined the effectiveness of PAA for fungus control on channel catfish *Ictalurus punctatus* eggs. The study consisted of five PAA concentrations (2.5, 5, 10, 15 and 20 mg/L) and an untreated control in a flow-through system. A single spawn was used for each replication (N=4). Eggs were treated twice daily until the embryos developed eyes. When hatching was complete for all viable eggs, fry were counted to determine the percent survival in each treatment. Fungal growth was severe in the untreated controls resulting in 11% survival. Treatments of 2.5, 5 and 10 mg/L PAA were significantly different from the controls ($P < 0.05$). The highest percent survival of hatched fry, was 5 mg/L PAA administered twice daily; the 2.5 mg/L PAA treatment had slightly less survival, but gives a higher margin of safety in case of treatment error. Very little fungus was present in treatments receiving 2.5 mg/L PAA or higher, and concentrations of 15 and 20 mg/L PAA were toxic to the eggs. The mean survival in the 0, 2.5, 5, 10, 15 and 20 mg/L PAA treatments were 11, 60, 63, 62, 32 and 0%, respectively. Therefore, PAA may be a novel compound which merits further investigations regarding its use in U.S. aquaculture.

Disinfection with peracetic acid (PAA), an alternative against fish pathogens

Thomas Meinelt[a], Lars-Flemming Pedersen[b], David L. Straus[c], Dibo Liu[a]

[a]: Leibniz-Institute of Freshwater Ecology and Inland Fisheries, Mueggelseedamm 301, 12587 Berlin, Germany. Email: liu@igb-berlin.de, meinelt@igb-berlin.de

[b]: DTU-Aqua, National Institute of Aquatic Sciences, Section for Aquaculture, North Sea Research Centre, P.O. Box 101, DK-9850 Hirtshals, Denmark. Email: lfp@aqua.dtu.dk

[c]: U.S. Department of Agriculture, Agricultural Research Service, Harry K. Dupree—Stuttgart National Aquaculture Research Center, P.O. Box 1050, Stuttgart, AR 72160, USA. Email: dave.straus@ars.usda.gov

ABSTRACT

Because of the lack of approved substances to treat fish diseases, disinfecting substances are tested to treat fish pathogens. These agents should not leave dangerous residues in the environment in order to successfully contribute to sustainable aquaculture. One of these substances is peracetic acid (PAA). PAA is an agent used for disinfection in aquaculture, but it must be investigated thoroughly in order to mitigate diseases without harmful effects on fish. We studied the effectiveness of several commercial PAA products to reduce fish pathogens *in Vitro* and developed *in Situ* treatment strategies. The tested commercial PAA products showed different molecular PAA: H_2O_2 ratios. The fish pathogens tested *in Vitro* were the bacteria *Aeromonas salmonicida, Yersinia ruckeri* and *Flavobacterium columnare*; the fungus *Saprolegnia parasitica* and the protozoa *Ichthyophthirius multifiliis*. Effective concentrations and exposure times of PAA products were determined and compared. One PAA product was tested *in Situ* in a rainbow trout RAS to reduce losses caused by bacterial gill disease (BGD). With two-day applications of 2 ppm PAA, the colony forming units (CFUs) were reduced from 10^6 to 10^4. The clinical symptoms of the BGD and the losses of fish were reduced. Although the application was done in the raceways of the RAS without a disconnection of the filter, the nitrifying bacteria were not affected.

Identifying Effective Direct-Fed Probiotics in Aquaculture

David Drahos[1], Seth D'Imperio[1], David Kuhn[2], Dan Taylor[2]

[1] Novozymes Biologicals, Salem, VA, USA
[2]Department of Food Science and Technology, Virginia Tech, Blacksburg, VA, USA

ABSTRACT

The identification of effective probiotic bacterial strains in Aquaculture is dependent on reproducible *in-vivo* methods that provide clear correlation with growth, feed conversion and health. Certain specific spore-forming Bacillus strains have been found to germinate effectively in gut tissues of Tilapia, Pangasius, and Trout and significantly enhance microvilli length and density. In addition, assessment of 1,032 metabolites produced in the liver and plasma of treated fish, have identified a subset of at least eight novel bile acids that are induced by the presence of a specific direct-fed Bacillus strain. An increase in fillet weight of 8-11% in fish fed one of these Direct-Fed Bacillus (B. subtilis SB3086; NRRL B-50136) has been observed in Pangasius and Tilapia. However, no microvilli enhancement was observed for at least one other Bacillus strain (SB3189), and no fillet weight change was found compared to fish in the non-treated Control group. Germination of the Bacillus spores can be quantified using qPCR methods which differentiate spores from germinated (vegetative) cells in extracts from various fish gut segments. It was also possible to visualize Bacillus spore germination in the gut of a living animal (*L. vannemi* shrimp). Here, a fluorescent *gfp* protein was expressed *in-vivo* from a germination-specific promoter on a plasmid introduced into the Direct-Fed Bacillus strain. The spores of this strain were coated onto shrimp feed, then fed to the active shrimp. Within 5 hours, the glowing fluorescence (520nm) can be readily observed in the gut tube of the relatively clear carapace of this living animal under 480nm excitation-light.

Reducing the Costs Due to Bio-fouling in Aquaculture Heat Exchangers with I2 Vapor Infusion

Michael Radicone

ABSTRACT

Maintaining proper tank water temperature offers the aquaculture industry a number of benefits including; optimal growth rates, shorter time to market and increased production numbers. At all growth stages, the health of the fish is improved by providing an acceptable water temperature. Temperature determines which organisms will thrive and which will diminish in numbers and size. There is a range of temperature that produces optimal abundance. The effects of temperature upon life of a cold blooded are profound since their body temperatures follow closely the temperature of their medium. Higher water temperature causes a number of potentially lethal conditions; increased toxicity through enhanced solubility of toxic substances, decreased dissolved oxygen and higher than normal metabolic rate which puts additional stress on the fish. Land based aquaculture farms use systems inclusive of heat exchangers and chillers to maintain proper heat transfer and although the need to maintain proper water temperature is paramount, bio-fouling of these systems can quickly cause a lethal alteration in tank water temperature. The energy costs to maintain acceptable temperatures are quickly increased due to ineffective heat transfer and the cost of maintaining these systems is substantial.

The Problem

Aquaculture systems reduce capital and operational costs by using heat recovery through heat exchangers. The effluent of any thriving fish farm or hatchery is rich with nutrients that can induce bio-film formation within the service water side whereas cooling water from ponds, streams or seawater can quickly foul the cooling water side. In some cases, less than a week is required to completely choke off the water flow with significant preemptive system performance loss. The resulting bio-films and algal growth that form on heat exchange surfaces cause severe and costly problems such as:

- Heat transfer surface fouling
- Pressure drop
- Microbiological-induced corrosion
- Shutdown for cleaning

The reduction in system functionality, labor and chemical costs and the effect on stock extracts considerable dollars from an aquaculture facility. Typically, the treatment is weekly back flushing in conjunction with hyper-chlorination to inactivate the offending species and break down bio-films. Dosing cooling water systems with chlorine in various forms to prevent fouling has long been a common practice in commercial and marine applications. Although chlorine is widely used due to its cost and availability, the EPA and other environmental agencies around the world restrict levels in water effluent discharge streams because of its detrimental effect on life forms through the formation of disinfection byproducts.

I2 Vapor Perfusion (I2VP) is a patented technology whereby through bubble/microbe contact, a vaporous and aqueous transfer of elemental iodine is made from the bubble surface to the microbial cell wall, inactivating or even killing the microbe.
During infusion, I2VP bubbles present a small amount of dissolved iodine to the water at an estimated 30 – 100 ppb, providing a sufficient residual for further microbe deactivation. In the presence of organic materials, the dissolved iodine is neutralized and the killing occurs through an interaction between the bacteria and the I2 vapor contained in the bubbles.

University and Governmental Studies
Numerous university studies have confirmed the disinfectant qualities of iodine vapor infusion. Studies at Cornell University, University of Illinois, Los Alamos Labs and Washington State University showed that a direct 90-second infusion of I2VP inactivates microbes in fluids, in fluids containing a high organic load and within established bio-films. A study, performed as part of the NESDI program, compared iodine to chlorine using a standardized EPA toxicity test protocol. This study clearly demonstrated that compared to chlorine;

- iodine has a lower potential for toxicity but also a significant mediating effect post-LOEC
- iodine has a greater persistence and reduced organic matter interaction

I2 VP technology dramatically reduces fouling in land and marine heat exchangers as well as macro fouling in seawater environments. Two studies—an Environmental Security Technology Certification Program (ESTCP 2013 WP-201219) and a Navy Environmental Sustainability Development to Integration (NESDI Project 502) project—investigated the ability of iodine infusion to reduce fouling formation within U.S. Department of Defense (DoD) shipboard heat exchangers and submarine water conduits. The ESTCP study indicated that infusing a shell-and-tube heat exchanger for five minutes per hour per day increased the time between physical cleanings by 300%. Currently under study for barnacle retardation, I2 Vapor Perfusion has been shown to significantly reduce hard foul formation with no rise in background seawater iodine levels. Easily integrated, I2 Vapor Perfusion may offer the means by which to prevent bio-fouling and other microbial conditions within the aquaculture environment without excessive biocide residue and expense.

Molecular analyses the microbial community dynamics in three RAS with different salinities for production of European eel

Mohamed E. Megahed

Mohamed E. Megahed National Institute of Oceanography and Fisheries (NOF), Gulfs of Suez and Aqaba's branch, Suez 41522, Egypt E-mail: aquageimprove@gmail.com Mobile: +201204588809

ABSTRACT

Farmed eel lives in water with high loads of different bacterial communities. The microbial profile in the culture medium different from those eel lives in natural water bodies. The eel larvae and microbial communities are in continuous interaction. The present work demonstrated that stable microbial environments improve the health growth and survival of reared eel larvae. The present work aimed to understand the microbial communities dynamics and ecology in RAS system. Part of the work focus on study the optimal microbial conditions for both eel health and the efficient nitrification process in the bioreactor. In this study eels larvae (0.5 g ± 0.01) were grown in three RAS systems operated at different salinities (0, 15 and 35 ppt). The microbial community dynamics and the effects of salinity on health and growth performance of eel larvae, as well as water quality. Water samples were collected from inlet and outlet rearing tanks (triplicate sample from triplicate rearing tanks), inlet and outlet moving bed bioreactor water, and bioreactor biofilm carriers from each system at three time points spanning a 2 months periods. 16S rDNA amplicone/DGGE techniques were used to analyze bacterial communities. The results showed that there were no differences in the microbial communities within each RAS and the microbial communities of the rearing water were similar to those of the bioreactor water. The microbial communities of water going in to the bioreactor were similar to those of the water going out It was evident that the microbial communities were not influenced by the bioreactor treatment. However, the microbial communities were dynamic with respect to the salinity concentration and the sampling time. The biofilm microbial communities significantly differed from water microbial communities, but it was less dynamic than water communities. The biofilm communities were stable over sampling periods and less influenced by salinity.

Molecular Characterization of *Aeromonas Hydrophila* Strains Isolated From Diseased Marine And Freshwater Fish

Mohamed S. M. Gado[1], Nadia B. Mahfouz[1] Amira A. E. Z. M. Omer[1], and Tadashi Shimamoto[2]

1) Faculty of Veterinary Medicine, Kafrelsheikh University, Egypt

2) Laboratory of Food Microbiology and Hygiene, Graduate School of Biosphere Science, Hiroshima University

ABSTRACT

Molecular techniques, such as PCR, can be used to solve the fish problems and increase sensitivity and specificity of pathogen detection. The aim of the present study is to isolate and identify *A. hydrophila* from commercial fish and to analyze with respect to the prevalence of virulence genes. *A. hydrophila* were isolated from 150 commercial fish samples of different ten fish species in Higashi-Hiroshima, using Aeromonas medium base (RYAN), a selective diagnostic medium. These isolates were identified by PCR using *A. hydrophila*-specific 16S rRNA primers. Detection of ten virulence genes were carried out in the identified *A.hydrophila* isolates by PCR using specific primers. The virulence genes include Hidrolipase (*lip*), elastase (*ahyB*), cytotonic enterotoxin (*alt*), temperature-sensitive protease (*eprCAI*), serine protease (*ahp*), aerolysin (*aer*), lipase (*pla/lip*), cytotoxic enterotoxin (*act*), haemolysin (*hlyA*) and cytotonic heat stable enterotoxin (*ast*). PCR screening revealed 61 isolates positive for *A. hydrophila*-specific 16S rRNA and the presence of six virulence genes (aerolysin, lipase, cytotoxic enterotoxin, cytotonic heat stable enterotoxin, temperature-sensitive protease and haemolysin). It also showed that the majority of the examined isolates carried one or more virulence genes. Infection experiments with the *A. hydrophila* isolates in fish give us important information about the virulence factors in *A. hydrophila*.

Growth Performance And Nutrient Utilization Of *Clarias gariepinus* (African Catfish) Fingerlings Fed Varying Levels Of *Moringa oleifera* (Drumstick) Leaves And Seed

Okayi R.G. and Neku Deborah .J

Department of Fisheries and Aquaculture,
University of Agriculture, Makurdi, Benue state, Nigeria.

ABSTRACT

The study of growth performance of Clarias gariepinus fingerlings fed with Moringa Oleifera (Leaves and seed meal) conducted for eight (8) weeks. Seed and leaves meal was used to supplement protein at different levels of inclusion: 0% (control), 10% Moringa Seed Meal (MSM), 10% Moringa Leave Meal (MLM), 5% of MSM and 5% of MLM respectively. Experimental diets were fed to the fingerlings at 5% biomass twice daily in hapas of $1m^3$.The fish were stocked at 20 fingerlings per hapas of an initial mean weight of 25.68 ± 0.041. Proximate composition of Moringa, it is shown that the seed meal has high crude protein content more than the leave meal. At the end of the study, experimental fish fed with test diet (TD2) had the highest mean weight gain and significant difference occur between the treatment ($P<0.05$), specific growth rate (SGR) and feed conversion ratio among the treatment. Water quality parameters were within the acceptable range.

Some Bacterial and Fungal affections Causing Disease Problems in Cultured Seabream *Sparus Aurata*

Riad H. Khalil [1] ; T.T. Saad [1]; Samia El-Hoshy [2] and Eman M . Haraz [2]

[1] Poultry and Fish Diseases Dept., Faculty of Veterinary Medicine, Alexandria University, Egypt.
[2] Animal health research institute (AHRI), Agriculture Research Center, Egypt.

ABSTRACT

Seabream (Sparus aurata) is a marine fish with economic value and widespread all over the world especially in the Mediterranean Sea. Seabream culture was known recently in Egypt and need for progressive development especially in feeding and health care. So the present study was conducted to through alight spot on some diseases of seabream which interfering culturing of it specially bacterial and fungal diseases. A total number of 50 cage-cultured seabream (Sparus aurata) of different body weight range (50 g to 200 g) and 15 to 18 cm in length were collected showing clinical signs from private fish farm at Wadi-Mariut region at west Alexandria governorate, Egypt. And submitted to full clinical investigation, postmortem, bacteriological and mycological examination. Affected seabream clinically inactive, loss of balance and loss of escape reflex, sluggish swimming and swim near to the water surface, distended abdomen, focal or diffused hemorrhages on different parts of the body, uni or bilateral corneal opacity, loss of scales and pigmentation, erosion and ulceration in the lateral aspect of the body. Internally the affected seabream showed congested gills. Liver showed reddish necrotic foci occur on its surface, distended gall bladder, congested spleen and accumulation of bloody tinged exudates in the abdominal cavity with offensive odour. The bacteriological examination of affected seabream revealed the isolation of *Photobacterium damselae, Burkholderia mallei, Vibrio alginolyticus, Pseudomonas alcaligenes, Aeromonas salmonicida, Staphylococcus heamolyticus , Acintobacter heamolyticus , Burkholderia cepaica , Alloiococcus otitis* and *Shewanella puterfaciens* according to morphological, biochemical and VITEK2 COMPACT SYSTEM.

From the macroscopic and microscopic characters of isolated fungi, it cleared that the isolated fungi was Penicillium sp, Aspergillus flavus , Candida sp , Legenidium sp , Fusarium sp , Rhizopus sp , Aspergillus fumigatus , Rhodotorula and Alternaria . Penicillium sp was isolated in a total 36 isolates from total 99 isolates.

Case study of an indigenous recirculatory aquaculture system in a fish hatchery of Naihati, West Bengal, India using tilapia as biocontrol agent

Srijit Chakravarty, N.K. Chadha, S. Dasgupta, B.K. Mahapatra, Paromita B. Sawant, S.K. Das

ABSTRACT

The physicochemical and biological quality of water largely determines the success and thereby productivity in eco-hatcheries. Availability of good quality water in hatchery operation is becoming scarce day by day because of changes in precipitation pattern, depletion of ground water table and environmental pollution.

The usual practice of the hatchery owners of Naihati, North 24 Pgs., West Bengal, India is to use ground water to carry out the breeding of fishes like Indian Major Carps and *Pangusius* sp. in their eco-hatcheries. As such type of hatcheries need huge water (4000-7000 litres per minute), it is becoming necessary to recycle and reuse the water after one cycle of operation so as to make economic water budgeting through recycling.

This is a case study of a hatchery owner in Naihati, North 24 Pgs., West Bengal, India, who has developed a novel method of recirculating the waste water of the incubation tanks through a pond stocked with Tilapia (*Oreochromis mossambica*) and using the same water again in the subsequent breeding operations. The pond (40 ft. X 40 ft.) is stocked with 2000-3000 nos. tilapia fry and reared without any external feed input. The effluent water, containing metabolites and egg shells provides as excellent nutrition for the stocked fishes, and tilapia, being a voracious feeder as well as prolific breeder, cleans the water of such impurities. Also, because of primary production and agitation of the tilapia the dissolved oxygen content of water increases significantly and the water becomes naturally seasoned for the subsequent use in the hatchery. This technique has been operational since 20 years and has never led to any mishap or failure in hatchery operation. This technique is thus an economic, self-sustaining towards water budgeting and optimizing the production cost through eco-friendly management practice ideal for aquaculture in developing countries.

Effects of direct-fed probiotics on fish production, physiology, and the metabolome of fish

David Kuhn[1], Dave Drahos[2], Daniel Taylor[1], Seth D'Imperio[2], Stephen Smith[3]

[1] Department of Food Science and Technology, Virginia Tech, Blacksburg, VA, USA
[2] Novozymes Biologicals, Salem, VA, USA
[3]Virginia-Maryland Regional College of Veterinary Medicine, Virginia Tech, Blacksburg, VA, USA

ABSTRACT

Probiotics have been gaining attention by the aquaculture community in recent years because they can be used to improve the overall health and production of fish. More specifically, research has demonstrated that probiotic fed fish are often able to more efficiently utilize nutrients in diets and enhance disease resistance when exposed to pathogens. Two studies were conducted comparing the effects that direct fed microbes (*Bacillus subtilis* SB3086) have on two different species of cultured food fish, tilapia (*Oreochromis niloticus*) and pangasius (*Pangasius hypophthalmus*).

The tilapia study was conducted at Virginia Tech's Southwest Virginia Aquaculture Research Center (Saltville, Virginia, USA) in two identical and independent recirculating aquaculture systems each equipped with twelve 210 liter round polyethylene tanks, a bubble bead filter for mechanical filtration, a fluidized-bed bioreactor for biological wastewater treatment, a UV disinfection unit, immersion titanium heaters, and distributed diffusion aerators. The pangasius study was performed at the Virginia Tech Human and Agricultural Biosciences Building (Blacksburg, Virginia, USA) in four identical systems equipped equivalently to the tilapia trial, and modified for six 210 liter tanks per system. Treatments were separated by system to maintain biological control. Dependent variables included fish production performance variables, biometrics, germination of probiotic bacteria, gut health, and changes to the metabolome of fish.

Gut germination of probiotics in the intestinal tract of tilapia that were fed probiotics were verified using quantitative real-time polymerase chain reaction (qPCR). Spores and vegetative cells were quantitatively differentiated. No fish mortalities occurred during the 14 week trial. In summary, fish that were fed the probiotic diet exhibited significantly ($P < 0.05$) greater skin-on fillet yields (39.6% MR) compared to fish that were fed the control diet (35.6% MR). Transmission electron microscopy (Figures 1a, 1b) revealed that fish fed probiotics had significantly ($P < 0.05$) longer (28%) intestinal enterocyte microvilli length. Meanwhile, scanning electron microscopy revealed that the density of microvilli was also significantly greater ($P < 0.05$) by 19%. The longer length and increased density of intestinal enterocyte microvilli can improve nutrient absorption and reduce translocation of microbes and associated toxic compounds into the peripheral vasculature.

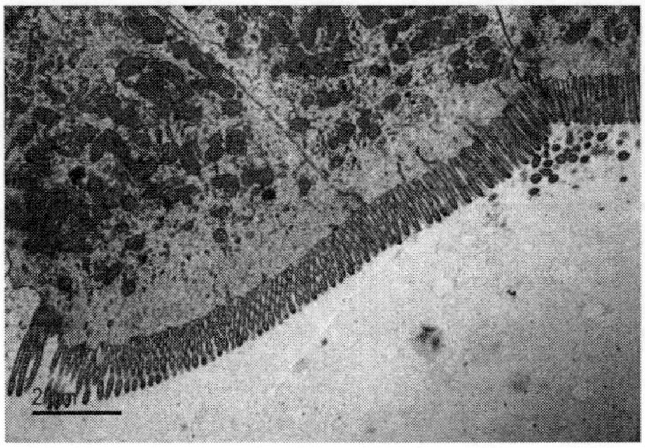
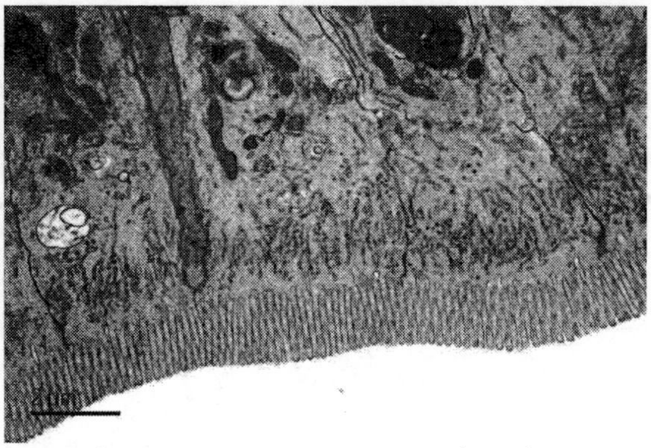

Fig. 1a – Control group (No probiotics) *Fig. 1b – Treatment group (Fed probiotic B1)*
Representative examples of intestinal microvilli under transmission electron microscopy.

The results from the pangasius trial were similar to the trial reported above for tilapia. Even though these enhancements are observed, very little is understood on how or if probiotics initiate biochemical changes at the cellular or systemic level. Fortunately, new laboratory-based tools are now available to empower researchers and the industry to answer these kinds of questions. One-hundred and ninety-two pangasius were used for this study. Half of the fish (n=96 were fed a control diet and the other half (n=96) were fed a probiotic diet (*Bacillus subtilis*). Fish were cultured on these diets for over 8 weeks in independent but identically designed recirculating aquaculture systems with similar water quality ($P > 0.05$) for dissolved oxygen, ammonia, nitrite, nitrate, pH, and alkalinity. After 8 weeks, half of each group of fish per diet were to *Edwardsiella ictaluri* via bath immersion at a sub-lethal level. After seven days, all fish with humanely euthanized and assessed for gross clinical signs of infection/disease in various tissues, and liver and plasma samples were collected for biochemical profiling.

For biochemical profiling, pooled samples of plasma and livers were quickly collected from euthanized fish and were frozen and stored at -80 C until analyzed. All samples were analyzed using a Waters ACQUITY ultra-performance liquid chromatography (UPLC) and a Thermo Scientific Q-Exactive high resolution/accurate mass spectrometer interfaced with a heated electrospray ionization (HESI-II) source and Orbitrap mass analyzer operated at 35,000 mass resolution. The present data set looked at 530 and 502 named biochemicals in the liver and plasma, respectively. Statistical differences ($P<0.05$), and approaching significant differences ($0.05<P<0.10$), were observed for biochemicals between control- versus probiotic-fed fish that were either exposed to or not exposed to *E. ictaluri*. Many notable shifts in biochemical were noted between treatment groups, thereby indicating that diet, infection, and diet X infection all had an impact on the metabolome of fish. Some of the notable findings from this study were: bile acids were both sensitive to both diet and infection (i.e. provides insight on how bacterial co-metabolize in the intestine), infection of probiotic fed fish elevated liver polyunsaturated fats (i.e. polyunsaturated fats modulate the immune system), infection induced plasma endocannabinoids are more robust in probiotic fed fish (i.e. endocannabinoids play a role as signaling molecules for immune function), and branched-chain amino acids and catabolites were elevated in fish fed probiotic diets (i.e. increased levels of these metabolites may indicate that probiotic fish are better able to synthesize glucose and fatty acids). These findings and other results not reported will be presented in more detail.

Application of Direct-Fed Probiotics to Improve Survival in Shrimp and Reduce Pathogenicity of EMS/AHPND Bacteria

Daniel Taylor[1], Ann M. Stevens[2], Moonyoung Choi[1], David Drahos[3], Seth D'Imperio[3], David Kuhn[1]

[1] Department of Food Science and Technology, Virginia Tech, Blacksburg, VA, USA
[2] Department of Biological Sciences, Virginia Tech, Blacksburg, VA, USA
[3] Novozymes Biologicals, Salem, VA, USA

ABSTRACT

Diseases affecting shrimp contribute to billions of dollars of economic loss yearly to the global aquaculture industry. *Vibrio parahaemolyticus* is an emerging CDC-reportable bacterial pathogen long associated with human seafoodborne gastroenteritis. However, in 2009, a new strain causing Early Mortality Syndrome (EMS) or acute hepatopancreatic necrosis disease (AHPND) in shrimp emerged with a devastating and ongoing impact on global shrimp production.

Probiotics have become an attractive method of pathogen control in shrimp culture. Probiotics act as natural immune enhancers and promote disease resistance in shrimp, while hindering pathogenic bacteria through competition for space and nutrients among other mechanisms. Shrimp losses attributed to bacterial disease can be greatly reduced through disease outbreak prevention. The overall project goal is to improve intensive shrimp production through direct application of probiotics in aquaculture feeds. It is hypothesized that direct-fed probiotics will colonize in the shrimp gut, enhance nutrient utilization, and provide increased disease resistance against pathogenic bacteria. Accordingly, we conducted the following studies to: (1) confirm the germination of direct-fed probiotic *Bacillus subtilis* spores in the shrimp gut, (2) evaluate how to infect shrimp with a *V. parahaemolyticus* strain of bacteria that causes AHPND in shrimp (*L. vannamei*) using controlled-challenge studies, and (3) determine the effectiveness of probiotics, containing *Bacillus subtilis,* for improving AHPND disease resistance.

The ability of *Bacillus subtilis* spores to germinate within the shrimp gut was confirmed by qPCR. Copy numbers of vegetatively growing cells were determined based on extrapolation against a standard curve.

The pathogenicity of three *Vibrio* strains was examined and it was concluded the *V. parahaemolyticus* strain identified as the causative agent of AHPND was the most lethal, based on bacterial challenge assays. Depending on dose, AHPND-infected shrimp exhibited 100% acute mortality within 24 hours of inoculation through feed. Shrimp were fed 3% their body weight during a four-day toxicity experiment with an application of 18-hour grown bacterial culture in equal volume to the feed weight. The number of bacterial cells added to the feed directly correlated with pathogenicity and only cells, not spent culture supernatant, was capable of causing death. Histological analysis of samples collected of shrimp killed by the AHPND strain revealed that there was a lack of central hepatopancreatic B-cells and massive sloughing of central tubule epithelial cells compared to shrimp not exposed to AHPND. These results strongly suggest degenerative pathology.

Bacterial challenge assays that were conducted to evaluate the benefits of direct-fed probiotics to minimize pathogenicity of the AHPND strain in shrimp demonstrated that particular strains of *Bacillus subtilis* were capable of significantly ($P<0.05$) reducing mortalities of shrimp that were repeatedly exposed to the AHPND strain of *V. parahaemolyticus* (Figure 1) in comparison to the absence of the probiotics strains. Closer examination of one strain (O14VRQ) resulted in consistently higher survival rates than control groups following AHPND exposure (Figure 2). The commercial application of specific strains of direct-fed probiotics at shrimp farms could help reduce disease outbreaks of AHPND. Additionally, these studies have helped form a consistent screening battery for *in vivo* evaluation of pathogenicity and interaction with probiotics.

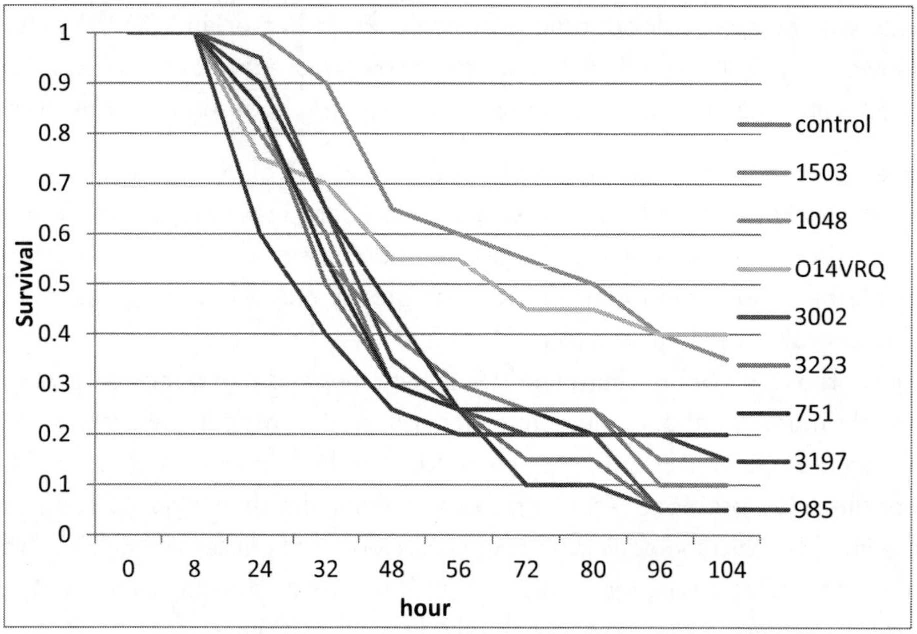

Figure 1: Multiple B. subtilus strain evaluation in AHPND disease exposure challenge

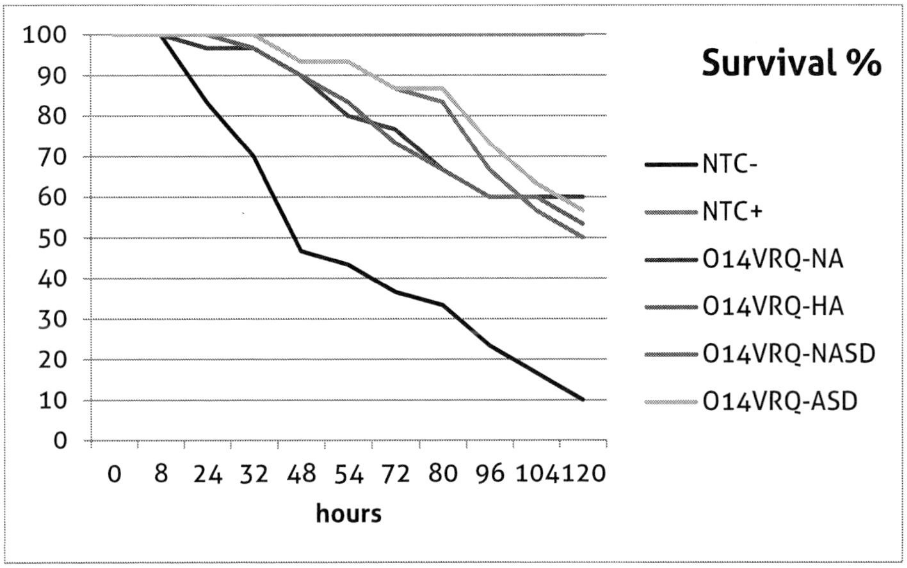

Figure 2: Bacillus strain O14VRQ application in AHPND disease exposure challenge

Application of Prebiotics in Recirculating Aquaculture Systems

Delbert M. Gatlin, III and Waldemar Rossi, Jr.

Corresponding author: Delbert M. Gatlin III
Department of Wildlife and Fisheries Sciences
Texas A&M University
College Station, TX 77843-2258
Phone: 979-847-9333; FAX: 979-845-4096
d-gatlin@tamu.edu

ABSTRACT

Several different feeding trials have been conducted at the Texas A&M University System Aquacultural Research and Teaching Facility to evaluate the effects of plant-protein-based diets and prebiotic supplements on the production performance and gut microbiota of juvenile hybrid striped bass (*Morone chrysops* x *M. saxatilis*) and red drum (*Sciaenops oellatus*) in both indoor, recirculating aquaculture systems and outdoor ponds under low-salinity (~ 5 g/L) conditions. A fishmeal (FM) - based (Special Select® menhaden FM) diet was used as the Reference in all feeding trials along with experimental diets containing a combination of soybean meal (SBM) and soy-protein concentrate (SPC) as the primary protein sources. These diets were fed for at least 8 weeks in the various trials. The additive effect of a pre-biotic (GroBiotic® -A, GBA) on the overall production performance and gut microbiota also was evaluated.

At the end of each feeding trial, production performance parameters were obtained and samples of intestinal contents were collected. The composition of gut microbiota populations was examined through denaturing gradient gel electrophoresis (DGGE), following amplification of bacterial 16S rDNA. The analysis of band pattern relatedness was determined with Molecular Analysis Fingerprinting software, using the Dice Similarity Coefficient (SC, expressed as a percentage) for comparisons between sample band patterns.

Limited differences in production performance of hybrid striped bass and red drum fed plant-protein-based diets were found. The results of DGGE analysis revealed changes in gut microbiota populations in the various trials under different culture conditions. In red drum, the composition of the microbiota population of ingesta significantly changed in response to plant-protein-based diets relative to the Reference (SC < 80%), and was variably affected by GBA inclusion in the SBM diet. In HSB, gut microbiota populations in the ingesta differed (SC < 80%) among treatments, while those in the mucosa were similar (80% ≤ SC ≤ 95%) between the Reference and SBM + GBA treatments and differed from the SBM treatment.

The results of these studies showed that the composition of gut microbiota populations of red drum and HSB can be affected by diet as well as by culture system. Detailed analysis of bacterial species within hybrid striped bass and red drum will be presented along with how they were altered due to environmental influences.

Around the World in 5 RAS Farms

Esteban JM Emparanza

Billund AquakulturService AS
Montanavej 2
Billund 7190
Denmark

ABSTRACT

The quality of the water source that is used to grow the target species and the location of the farm itself are two of the most important factors shaping both the construction and operation milestones of a RAS system. Depending on the conditions of the water source, being this a lake, river, reservoir or borehole, it may be necessary to have a pre-treatment system to help obtain good quality water for growing aquatic species. Depending on the location, the construction will have different challenges to overcome and so to become functional. The following is a collection of anecdotes from five RAS projects built at different locations around the world. The main challenges, problem-solving situations, relevant water quality parameters of the water source and general outcome are depicted.

Zinc mine turned out to be a Gold mine. Located near the town of Daniel´s Harbour, in the northern peninsula of Newfoundland, Canada, this salmon farm has one of the most perfect raw water qualities ever. Isolated in the woods, the area is frequented by moose, deer, black bear and caribou. Mixing the water coming from an old Zinc mine, which is high in Zinc but also in alkalinity and hardness, with the water from a nearby bore hole, the mix produces a perfect water to grow fish, no lime needed to buffer the already well balanced water. The natural condition of the mixed water shows alkalinity levels around 300-400 mg/L and hardness levels around 400-500 mg/L. The Zinc levels showed to decrease to < 0.1 mg/L after mixing. Considering 3 meters of snowfall in just 3 days, this is really been "snowed under" with work. **Wild thing**. Lonnavale is located in the heart of Tasmania, the pristine, wild and iconic island south to Australia. The start-up of this plant was not exempt of trouble, being the raw water quality coming from a bore hole the main issue. Limestone soil presented us with 90 to 100 mg/L of CO_2 coming in the water, along with elevated levels of Iron (>5 mg/L) and Magnesium (>20 mg/L). A pre-treatment system for the raw water needed to be built. Despite that and some wild things happening during construction of the site, like a small Wallaby jumping to have a swim in the degassing filter sump and not one but two highly dangerous brown snakes found sneaking in the surroundings, things ran smoothly afterwards. **Silk Road´s newest star**. EJRAS 500 is located deep in the north west of China, just in the outskirts of the Gobi desert. What do Bactrian camel and Rainbow trout have in common? They grow in one of the harshest environments a RAS farm has been placed. Extreme environmental temperatures, ranging from -35°C in winter and +37°C in summer, meant that maintaining the stable water temperature the fish need to grow at, has been quite a challenge. However, once this main challenge had been addressed, the farm is on its way to produce a first in the region, 5-kilogram Rainbow trout in a land-based system. **Fish spa.** Jurassic Salmon in North West Poland, also know as West Pomerania, is a unique farm using geo-thermal water coming from 1,200 m underground for the production of Atlantic Salmon in their land-based Grow-out system. The source of the geo-thermal water was located 5 kilometres away from the farm, so proper piping had to be performed to carry the water to the site. The pristine and superb conditions of this water that comes naturally at 18°C and 14 ppt salinity, water that was originally thought to be destined for a Spa, make the fish grown here produce a very fine fillet, which will make its debut on the market very soon. **Viking salmon**. Belsvik is placed in a quiet and beautiful shore west to the Trondheim area, Norway. Smolts are grown using water from a nearby lake, which comes with relatively high TOC levels that protect the fish from any waterborne heavy metals. The modern building can be walked through a long corridor, looking at the fish being grown at one side and filtration systems operating to the opposite side through proper windows. Chilly temperatures made icy roads and paths that became relatively dangerous under construction, not a challenge for a Nordic, but difficult enough for the aliens. When the smolts are ready for transfer to sea, they are directly sent by piping to the wellboat waiting anchored in the farm´s own dock.

A growth model for gilthead sea bream (*Sparus aurata* L.)

Ido Seginer

Civil and Environmental Engineering, Technion, Haifa 32000, Israel

ABSTRACT

Gilthead sea bream (*Sparus aurata* L.) is a fish commonly cultivated around the Mediterranean sea and elsewhere in marine cages and recirculating aquaculture systems. Managing these facilities requires a growth model which describes the response of fish to their environment. The dominant predictors of growth rate, assuming adequate water quality, are fish size, M, water temperature, T, and feed ration, F. Over the past 30 years many experimental studies with gilthead sea bream have been conducted, each providing partial information regarding the general growth function $G\{M,T,F,...\}$.

The simplest model to describe the growth of an organism over its natural life span is the von Bertalanffy equation. For fish it was found to be $G=\eta M^{2/3}-\kappa M$, where the first term represents anabolism and the second – catabolism. The coefficients η and κ represent the growth potential of the habitat, in our case the effects of temperature, T, and food availability, F. It turns out that the correlation between the anabolic and catabolic terms under Mediterranean aquaculture conditions is high enough to allow the simplification of the two-term growth equation to a single-term equation of the form $G=\zeta M^{\nu}$.

The response of growth to temperature is a unimodal function, with a maximum at about $T_{\mathrm{opt}}=26$ °C. In Mediterranean aquaculture water temperature is often sub-optimal for gilthead sea bream, dropping sometimes below the temperature at which the fish stop growing, $T_G \cong 12$ °C. In the sub-optimal temperature range the response to temperature is approximately linear, leading to the currently popular Thermal-Growth-Coefficient (TGC) model $G=p(T-T_G)M^{\nu}$, where p may be formulized as a function of the feed ration.

The response to feed may also be approximated by a linear relationship, starting at zero growth when feeding covers just the needs of maintenance, and reaching a maximum at feeding to satiation. Introducing this approximation into the TGC model, the result is $G=p_s\left(\frac{r-r_m}{1-r_m}\right)(T-T_G)M^{\nu}$, where p_s is p at satiation, r is fraction of satiation feeding, $r \equiv F / F_s$, and r_m is r at maintenance feeding. The satiation feed ration, F_s, may be approximated by $F_s = q_s(T - T_F)M^{\alpha}$, which concludes the formulation of the model.

The linear temperature response is biologically inappropriate, but within the bounds of Mediterranean aquaculture (short life span and sub-optimal temperature), and in view of the wide scatter of data, it may be a useful model for practical purposes. The model may be adapted on-line for each particular growing facility, mainly by adjusting the parameters p_s and q_s.

Calibration of the model against available gilthead seabream data is summarized in the following three tables (for G in g/(fish · day), M in g/fish, T in °C and F in g[feed]/ (fish · day)).

Feeding to satiation model $F_s = q_s(T - T_F)M^{\alpha}$

Study	Author	Year	Country	Info type	Exp system	Size range g/fish	Temp range °C	q_s	T_F	α	Dev %	Ration 200g/20C g/fishd
1	Kraljevic	1984	Yugoslavia	Data	Tanks	2 - 110	12 - 23	0.0125	10.5	0.67	25.8	4.09
2	Petridis	1996	Greece	Data	Cages	2 - 350	14 - 24	0.0113	11.1	0.69	65.1	3.85
3	Andrade	1996	Portugal	Data	Pond	70 - 250	17 - 27	0.0075	7.6	0.68	25.0	3.32
4	Jauralde	2013	Spain	Data	Tanks	20 - 420	22 - 25	0.0046	11.6	0.72	23.5	1.70
	Mean							0.0090	10.2	0.69	34.9	3.24
5	Lupatsch	1998	Israel	Model	Tanks	1 - 400	20 - 26	0.0065	11.8	0.70	6.4	2.17
6	Lupatsch	2003	Israel	Model	Tanks	1 - 450	20 - 28	0.0055	11.0	0.69	10.5	1.95
7	Libralato	2008	Various	Model	Various	10 - 250	14 - 26	0.0056	12.1	0.71	13.1	1.91
8	Llorente	2013	Mediterranean	Table	Cages	10 - 500	13 - 27	0.0080	10.7	0.61	17.5	1.85
	Mean							0.0064	11.4	0.68	11.9	1.97

Growth model $G_s = p_s(T - T_G)M^{\nu}$

Study	Author	Year	Country	Info type	Exp system	Size range g/fish	Temp range °C	p_s	T_G	ν	Dev %	Growth 200g/20C g/fishd
1	Kraljevic	1984	Yugoslavia	Data	Tanks	2 - 110	12 - 23	0.0018	10.8	0.66	40.9	0.53
2	Petridis	1996	Greece	Data	Cages	2 - 350	14 - 24	0.0060	12.6	0.66	25.1	1.51
3	Brigolin	2010	Italy	Data	Cages	20 - 320	13 - 25	0.0071	11.4	0.60	23.3	1.43
4	Jauralde	2013	Spain	Data	Tanks	20 - 420	22 - 25	0.0026	11.2	0.69	30.1	0.89
5	Mayer	08,12	Spain	Data	Cages	6 - 460	13 - 27	0.0052	11.9	0.64	**	1.25
	Mean							0.0045	11.6	0.65	29.9	1.12
6	Lupatsch	1998	Israel	Model	Tanks	1 - 400	20 - 26	0.0035	11.5	0.70	9.7	1.22
7	Lupatsch	2003	Israel	Model	Tanks	1 - 450	20 - 28	0.0031	11.1	0.69	16.4	1.09
8	Libralato	2008	Italy	Model	Various	10 - 250	14 - 26	0.0030	13.9	0.71	39.8	0.78
9	Hernandez	2003	Spain	Model	Cages	10 - 1000	18 - 23	0.0024	12.1	0.69	42.8	0.70
10	Piedecausa	2010	Spain	Model	Cages	20 - 480	13 - 28	0.0030	7.6	0.64	21.9	1.08
	Mean							0.0030	11.2	0.68	26.1	0.97

Study	Author	Country	Year	Info type	Feed units	Size range g/fish	Temp range °C	r_m -
1	Klaoudatos	Greece	1986	Data	Feed	70-87	25	0.28
2	Klaoudatos	Greece	1996	Data	Feed	2-4	18	0.24
3	Lupatsch	Israel	1998	Data	Feed	25-130	23	0.17
4	Lupatsch	Israel	2003	Data	Energy	40-450	21	0.19
5a	Jauralde	Spain	2013	Data	Feed	20-420	21-25	0.15
	Mean							0.21

Noteworthy points are:

1 The experimental error (Dev) is quite large.

2 Nevertheless, the values of T_F, T_G, α, ν, and r_m are rather consistent among studies.

3 The values of the exponent of M (α and ν) are similar to the traditional 2/3.

4 The values of the threshold temperatures (T_F and T_G) are similar to the observed 12°C.

5 The parameters q_s and p_s vary somewhat more, but may be adapted on-line.

6 The predicted growth rate of a 200 g fish at 20 °C is about 1 g/(fish · day).

7 The best fit to the data yields the growth function

$$G\{M,T,F\}=\left(\frac{F/\left(0.007(T-10.8)M^{0.69}\right)-0.21}{1-0.21}\right)0.004\,(T-11.4)M^{0.67}\ .$$

Full-Scale CFD Modeling and Experimental Investigation Of Large RAS Tank Hydrodynamics

JMR Gorle[1], BF Terjesen[1], S Summerfelt[2]

[1]Nofima AS, Sjølseng, NO-6600 Sunndalsøra, Norway
[2]The Conservation Fund Freshwater Institute, 1098 Turner Road, Shepherdstown, WV 25443, USA

ABSTRACT

Recirculating Aquaculture System (RAS), unlike conventional methods, provides a unique set of farming conditions that favor the fish welfare and operational performance. There is currently a considerable interest in land-based and sea-based closed-containment aquaculture systems for salmon post-smolts (Terjesen et al., 2013a), with tanks up to several thousand m^3 volumes. Such systems are complex (Terjesen et al., 2013b), and complete experimental investigation of tank hydrodynamics is not possible. This paper presents a finite-volume based full-scale computational study, including the methodology development, of an octagonal Recirculating Aquaculture System tank of $788m^3$ for post-smolt production.

As shown in Figure 1(a), the post-smolt production tank has 2 inlet pipes each having 12 inflow holes that supply the working fluid tangentially into the tank. The outlet is located at the center of the tank. For the sake of simplicity, the computational domain excludes the structures such as external piping, walkway etc., which are not involved in the flow environment inside the tank. The structured computational grid as depicted in Figure 1(b) ensures an affordable CPU effort to solve the flow conservation equations using Realizable k-ε turbulence model. Several assumptions are made and defended for the model development to be feasible. Preliminary validation of CFD results is performed by comparing the velocity profile at the central plane along the tank's walkway against experimental measurements.

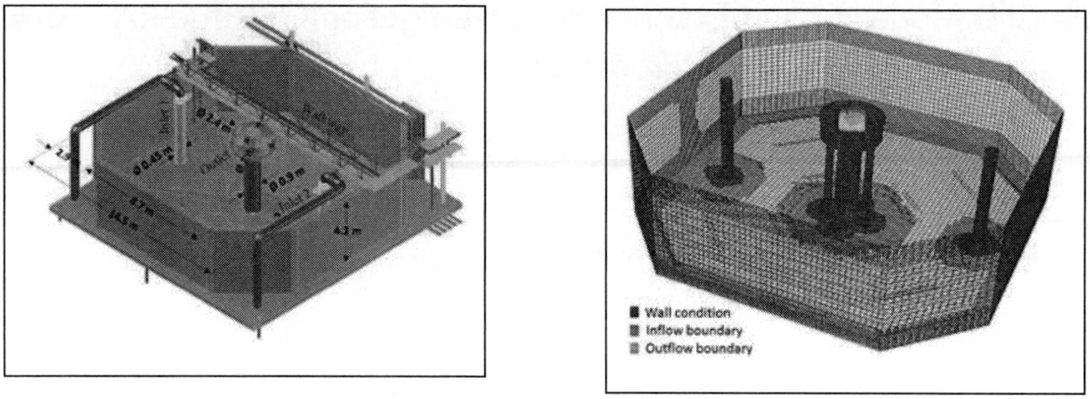

(a) CAD model of post-smolt RAS tank

(b) Computational mesh

Figure 1: Case under study. (a) Full-scale model with dimensions, and (b) structured grid with basic boundary conditions for CFD analysis

The objectives of this study are two-fold.

1. Develop a turbulence model to characterize the hydrodynamics of large culture tank: This involved the pre-processing, physics set-up, parallel processing and analysis.

2. Evaluation of key flow structures that influence the water quality.

The quantitative and qualitative observations of this CFD study opens the doors of the development of scale-models, which would reduce the analysis efforts. In addition, the present study will lead to the future studies on free-surface and vortex dynamics, which are critically important hydrodynamic behavior of large culture tanks.

References

1. Terjesen, B.F., Rosten, T.W., Ulgenes, Y., Henriksen, K., Aarhus, I.J., Winther, U., 2013a. Betydning av vannmiljøet ved produksjon av laksefisk i lukkede systemer i sjø. Water quality requirements for efficient farming of Atlantic salmon in closed systems. In Norwegian, English abstract. VANN 48, 14-27.

2. Terjesen, B.F., Summerfelt, S. T., Nerland, S., Ulgenes, Y., Fjæra, S.O., Reiten, B.K.M., Selset, R., Kolarevic, J., Brunsvik, P., Bæverfjord, G., Takle, H., Kittelsen, A.H., Åsgård, T. 2013b. Design, dimensioning, and performance of a research facility for studies on the requirements of fish in RAS environments. Aquacultural Engineering 54, 49–63.

The Effect Of Carbon Dioxide On Metabolism And Growth In Adult Pikeperch (*Sander lucioperca*)

K. Steinberg*[1,2], J. Zimmermann[1,2], K. Stiller[1,3] S. Meyer[4] and C. Schulz[1,2]

[1]Gesellschaft für marine Aquakultur mbH Büsum, Germany
Email: steinberg@gma-buesum.de
[2]Institute of Animal Breeding and Husbandry, Marine Aquaculture, Christian-Albrechts-University, Kiel, Germany
[3]Research and Technology Centre, West Coast, Büsum, Germany
[4]Chamber of Agriculture Schleswig-Holstein, Germany

Introduction

Partial pressure of CO_2 (pCO_2) in open water bodies is roughly that of the atmosphere but fish grown in RAS systems may be exposed to substantially higher carbon dioxide concentrations and it is one of the limiting factors of land-based aquaculture. Most studies of the impact of hypercapnia on fish have focused on acute hypercapnia. Only little is known on the chronic effects. Carbon dioxide has direct physiological effects on the fish as it increases plasma pCO_2 which reduces oxygen transport. Additionally reduced growth and feed conversion efficiency as well as nephrocalcinosis have been observed in longterm exposure of carbon dioxide to rainbow trout and Atlantic Salmon (Fivelstad, 2013).

Materials and Methods

The trial was conducted in a recirculating aquaculture respirometer system (RARS) as described in detail by Stiller et al. (2013). The system consists of 10 tanks (volume of 250 L each), one of which is kept empty of fish. The RARS (4 m^3 volume) has its own water treatment system but was connected to the pump sum of another RAS system (40 m^3 volume), to ensure water quality parameters and efficient stripping of CO_2. The system consists of an online water chemistry analysis unit measuring oxygen, water temperature and pH, dissolved carbon dioxide and total ammonia nitrogen. Ten fish were stocked at nine respiratory tanks with an average weight of 251.9 g (190.0 g – 308.3 g) and fed daily until apparent satiation with a commercial diet (ALLER Metabolica). Food grade CO_2 was added to the medium (~15 mg L^{-1} CO_2) and high (~ 30 mg L^{-1} CO_2) hypercapnia level tanks using ceramic diffusers while three tanks were left without CO_2 ingassing to represent the low hypercapnia regime (~5 mg L^{-1} CO_2). Individual body wet weight [g] and length [cm] of three individuals per tank and wet fish group weight [kg] were determined at the start, middle and the end of the experiment. Blood was taken from the caudal vein of three fish per tank to evaluate haematocrit and measure blood pH. Additionally three fish per tank were randomly selected for the analysis of the whole body composition both at the beginning and the end of the experiment.

Results

Fish showed a decreasing trend in both the daily feed intake with a DFI of 0.93±0.05 % for the low hypercapnia regime, 0.98±0.12 % for the medium regime and 0.88±0.04 % for the high regime and the SGR of 0.89±0.09 %, 0.86±0.08 % and 0.78±0.06 % accordingly. FCR's increased from 1.05±0.05 (low) to 1.14±0.04 (medium) and 1.13±0.07 (high). The condition factor was similar through all groups and treatments with a mean CF of 0.87±0.13. Hematocrit showed no differences within treatments but increased linear over time from 29±2 % at the start of the experiment to 29±7 % after four weeks to 40±10 % and the end of the experiment. Evaluation of the metabolic oxygen consumption rate (MOCR) showed a significant difference between the low and high CO_2 group and a linear correlation between all groups on the first day. The MOCR and the MOCR per gram feed decreased lineary over the course of the experiment within all CO_2 groups. MOCR was lowest at all times for the high CO_2 group and highest for the low CO_2 group in the first and the last week of the experiment. The metabolic nitrogen production rate (MNPR) decreased lineary over the course of the experiment for all three CO_2 groups and the MNPR per gram fed decreased lineary for both the medium and the high CO_2 group over the course of the experiment.

Discussion and Conclusion

The fish were fed till apparent satiation for a limited amount of time (10 minutes). The data retrieved for SGR, DFI and FCR suggest a reduction in appetite in correlation with the CO_2 concentration. This effect was also described in other fish species such as turbot (Stiller et al., 2015) or Atlantic salmon (Fivelstad, 2013). Changes in haematocrit were not significantly different between the hypercapnia treatments but increased linear over time thus indicating the impact of the respirometry system (oxygen saturation slightly below 100%) overlaying the impact of the CO_2 dosing. The analysis of metabolic oxygen consumption and nitrogen production rates in this study indicate a reduction in metabolic activity of pikeperch with increasing CO_2 concentrations thus suggesting reduced growth rates. The research provides new data in terms of adaptive mechanisms of pikeperch to hypercapnia conditions and underlines the importance of CO_2 measurement and management in closed RAS systems. Furthermore the research provides detailed baseline data on the energy metabolism of pikeperch.

References

Fivelstad, S., 2013. Long-term carbon dioxide experiments with salmonids. Aquac. Eng. 53, 40–48. doi:10.1016/j.aquaeng.2012.11.006

Stiller, K.T., Moran, D., Vanselow, K.H., Marxen, K., Wuertz, S., Schulz, C., 2013. A novel respirometer for online detection of metabolites in aquaculture research: Evaluation and first applications. Aquac. Eng. 55, 23–31. doi:10.1016/j.aquaeng.2013.01.004

Stiller, K.T., Vanselow, K.H., Moran, D., Bojens, G., Voigt, W., Meyer, S., Schulz, C., 2015. The effect of carbon dioxide on growth and metabolism in juvenile turbot Scophthalmus maximus L. Aquaculture 444, 143–150. doi:10.1016/j.aquaculture.2015.04.001

Assessing the Cost of Tilapia Production in an Airlifted POLYGEYSER® RAS Facility

Marlon A. Greensword*, Ronald F. Malone

Department of Engineering Science
Louisiana State University
Baton Rouge, Louisiana, 70803-6405
Mgree15@lsu.edu; rmalone@lsu.edu;

ABSTRACT

A commercial scale airlifted PolyGeyser® RAS tilapia growout facility is designed to support an annual production of one million pounds. The direct normalized cost is computed based upon estimates derived from current commercial practice. The impact of changes in loan interest rates is also examined.

The facility is housed in 10 polyethylene greenhouses and 1 concrete office and storage building. Buildings are arranged in the order of production stages:

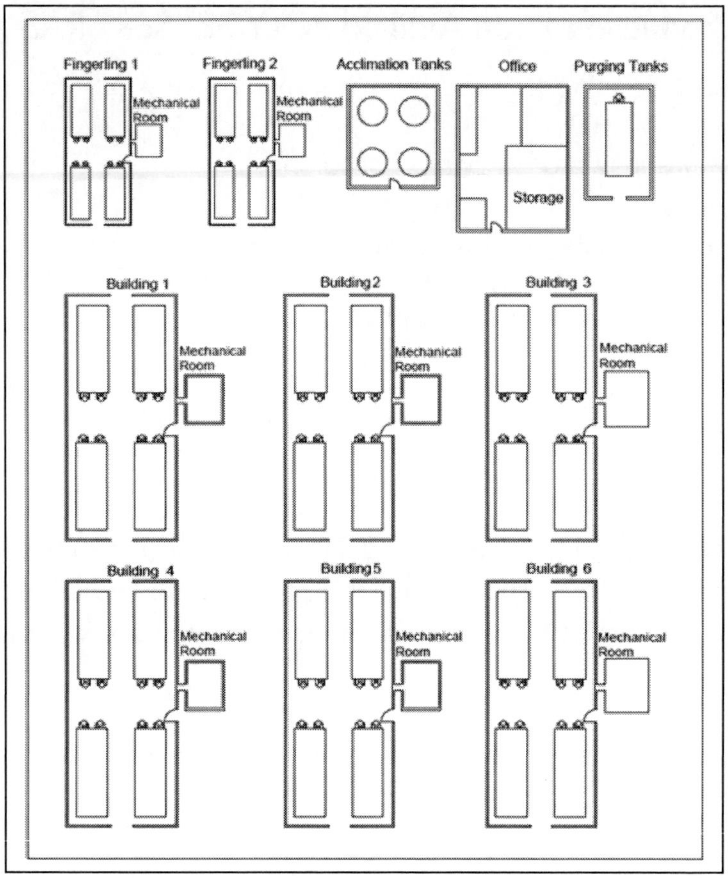

1 acclimation greenhouse for outsourced fingerlings, 2 fingerling buildings, 6 growout tank buildings, and 1 purge tank for pre-harvest fish. The facility is assumed to be equipped with a water well that serves as water source for all buildings. Fingerling buildings host four 30,000 gallon tanks each and the growout buildings each host four 45,000 gallon tanks. Each RAS was equipped with low profile PolyGeyser® bioclarifiers driven by airlifts. Fingerling tanks' airlift blowers have a size of 450 cfm, and their filters (PolyGeyser®) have 100 ft3 of beads. Growout tanks' airlifts and filters have a size of 675 cfm and 150 ft3 of beads, respectively. No water pumps were employed in the design. All tanks have one air blower pump. Those of fingerling tanks function at 450 cfm and 4 hp; those of growout tanks function at 675 cfm and 6 hp.

The total growout time is 42 weeks. Incoming fingerlings remain in an acclimation tank for one week, at the end of which they weigh 16 g each. They are then divided into equal amounts and transferred into the fingerling buildings. After 21 weeks, they each weigh 70 g and have reached a body length of 2.5 cm or more. On week 21, the content of each tank is divided into 2 equal amounts and placed into 2 growout tanks. The fish stay there for 21 weeks, at the end of which weigh above 850 g. They are then transferred to the purging area where they excrete remaining waste for one week. To maximize production, each tank remains operational on a constant basis, with the exception of 2 weeks down time every 350 days. Acclimation tanks are replenished every 4 weeks to ensure that a fingerling load is ready for growout transfer upon harvest. Fingerling and growout schedules must overlap for operation to remain constant. A growout cycle from 70 g to 850 g lasts 21 weeks. The facility produces 2 entire harvests (plus 65% of a third growout cycle) annually. The assumed facility amount 1,029,807.69 tilapia at harvest.

Results show a 30-year life cycle cost of $32,216,000 for the facility, $11, 056,000 (43%) of which is attributed to feeding. Operation represents 20% with a cost of $5,053,000. Equipment accounts for 15%. The stocking cost is $2,373,000 (9%) and the chemical cost is $2,341,000.00 (9%). The operation cost and equipment cost are $5,053,000 and $3,788,000, respectively. The building and maintenance costs (4%) follow at $535,000, and $426,000. Findings demonstrate the financial feasibility of airlifted PolyGeyser®-equipped facilities for domestically produced tilapia. The tilapia production cost of the proposed facility is $$1.19/lb. produced before processing.

NRCS Planning Considerations for Waste Management Systems Available for Recirculating Aquaculture Producers

Bill Reck, PE

USDA NRCS National Environmental Engineer, Natural Resources Conservation Service
United States Department of Agriculture
1400 Independence Ave, SW
Room 6130-S
Washington, DC 20250
(202) 720-4485

ABSTRACT

USDA Natural Resources Conservation Service provides financial and technical assistance to agricultural producers on a voluntary basis to plan and implement conservation practices that improve soil, water, plant, animal, air and related natural resources on agricultural land and non-industrial private forestland. Owners of land in agricultural or forest production or persons who are engaged in livestock (includes aquaculture), agricultural, or forest production on eligible land and that have a natural resource concern on that land may apply to participate in the Environmental Quality Incentives Program (EQIP). EQIP provides financial and technical assistance to agricultural
producers in order to address natural resource concerns and deliver environmental benefits, such as improved water and air quality, conserved ground water and surface water, reduced soil erosion and sedimentation or improved or created wildlife habitat.

NRCS can assist aquaculture producers who meet eligibility requirements with both technical and financial assistance for the design and installation of practices which provide environmental benefits such as waste storage structures, waste treatment lagoons, animal mortality treatment facilities, constructed wetlands for water treatment, waste transfer via pumping and pipes, waste solid/liquid separation, waste application via irrigation, and nutrient management of cropland receiving wastewater.

NRCS has a long history of providing technical assistance to the livestock industry and to a lesser extent, the aquaculture industry. Over the years, NRCS has developed tools for the planning and sizing of the components of a waste management system. A critical component of water quality conservation is proper management of animal waste. One of the key factors of animal waste management is the design of one or more

storage structures (ponds, tanks, and/or dry stacks) that can store the waste generated for time periods recommended by the state and local regulatory agency. An example of a tool developed by NRCS is the Animal Waste Management (AWM) software which was developed
and supported by the USDA/NRCS to assist in the analysis and design of these facilities. AWM is a planning/sizing tool for waste storage structures for animal feeding operations. It estimates the amount of manure, bedding, and process water based upon the
type and number of animals and management followed. This information along with the local climatic factors and other characteristics feed into the design of these facilities. The facilities that can be designed and evaluated by AWM include storage ponds, tanks and anaerobic lagoons for liquids, and covered and uncovered dry stacks with or without walls for solids.

Additionally, NRCS administers a Conservation Innovation Grants (CIG) program which are competitive grants that stimulate the development and adoption of innovative approaches and technologies for conservation on agricultural lands. CIG uses Environmental Quality Incentives Program (EQIP) funds to award competitive grants to non-federal governmental
or nongovernmental organizations, American Indian tribes, or individuals. Producers involved in CIG-funded projects must be EQIP eligible. Aquaculture projects in the past have included innovative waste management techniques like using geosynthetic tube to filter aquaculture waste solids.

This presentation will cover of NRCS technical and financial available assistance as well as an overview of planning/design considerations for recirculated aquaculture waste management and provide insight on the AWM software and its application for fish sludge or processing wastes.

Pneumatic Sludge Handling in Low Profile PolyGeyser® Floating Bead Filters

Ron Malone, Rhine Perrin and Tim Pfeiffer

Aquaculture Systems Technologies
108 Industrial Avenue
New Orleans, LA
Ron.Malone@astfilters.com

ABSTRACT

Sludge handling has occasionally proven problematic in both commercial RAS production systems and even in ornamental koi pond applications filtered by PolyGeyser filters. In contrast to propeller and bubble-washed units, that remove sludge at the time of backwashing, the PolyGeyser filters have the ability to store sludge across several backwash cycles. Manual sludge removal is required periodically, a task that is easily neglected.

In RAS applications, the frequency of sludge removal is driven by the feed application rate in kg feed/m^3 bead-day (lbs feed /ft^3 beads-day). Peak feed loading conditions which can exceed 32 kg/m^3-day (2 lbs/ft3-day) in commercial tilapia growout, generate solid waste on a dry weight basis around 0.25 kg/kg-feed consumed. This solid waste consists of a blend of partially digested fecal matter and dislodged biofilm. Sludge digestion in the bead bed is very dynamic and the exact amount of wastes produced are dependent on the temperature, loading, and backwash frequency employed. Assuming the mean sludge density in the sludge storage compartment is about 4 percent, the internal rate of sludge accumulation can be estimated at 6 liters/kg-feed (0.75 gal/#-feed); a rate which will fill the internal sludge storage capacity of a typical commercial scale PolyGeyser in about 1 day at peak loading. These units are usually backwashed 8-12 times daily with sludge removal once or twice a day. This frequent backwashing coupled with careful manual sludge removal yields a generation rate from the filter of about 8 liters/kg-feed (1 gal/#feed).

A rather menial task in a facility with 30-40 large bioclarifiers, early commercial practice showed it is often a task left undone, or partially done. A situation that leads eventually to a "dirty" backwash and consequently a turbid production tank that begins to act as a hybrid suspended growth/fixed film reactor system. To avoid this situation, all the early airlifted PolyGeyser facilities converted to airlift assisted sludge removal employing a backwash sensor, a delay timer, and actuator valve. The sludge removing airlift was activated for about 20 seconds about 60 seconds after completion of each backwash event. This approach successfully resolved the issue albeit with some increase in water loss to perhaps 32 liters/kg-feed (4 gal/#-feed).

In response, the new generation of airlift compatible or low profile PolyGeysers are designed to employ the pressure differentials produced during a backflush event to pneumatically move sludge out of the sludge settling compartment. Implemented without moving parts or electronics, these approaches can potentially alleviate sludge accumulation problems.

The most direct approach is the "Direct Pneumatic Discharge" (DPD) that results with the immediate removal of sludge during each backwash sequence. In this approach, the air pressure in the charge chamber is combined with the hydrostatic pressure of the sludge in an internal sludge storage basin to develop a controlled lift (typically about 25 centimeters (ten inches) just before a backwash event. This strategy produces a small amount of relatively concentrated sludge (4-6%) during each backwash cycle. DPD units are capable of operating literally for months without manual intervention, as long as air is continually delivered to the charge chamber. This strategy has proven sensitive to prolonged power interruptions that allow excessive thickening of sludge (>8 %) which can lead to clogging of the sludge discharge line.

Particulate effluent capture and utilization from recirculating aquaculture systems: Enabling integrated agricultural production systems

Todd C. Guerdat* [a], Thomas N. Reynolds [a]

[a] University of New Hampshire
* todd.guerdat@unh.edu

ABSTRACT

Over half of the world's seafood is produced from aquaculture. Eighty percent of the seafood we eat here in the US is imported resulting in nearly an $11 billion trade deficit for seafood alone. However, commercial aquaculture development in the United States has been slow primarily due to excessive site costs when locating close to target markets and effluent treatment requirements. The pressure on the natural fisheries increases every year while yields are decreasing due to overfishing and widespread environmental degradation. The demand for fisheries products by consumers continues to increase due to a higher demand by both a growing population and a more affluent and health conscience consumer. It should come as no surprise then, that aquaculture is the fastest growing food-producing segment of agriculture worldwide. However, to support the development of commercial aquaculture in the United States, a more economically sustainable means for producing seafood must be developed.

Aquaculture is agriculture, and agricultural systems may achieve economically sustainable production by integrating with other production systems. Integrated farming systems (IFS) improve energy and resource utilization, offer an opportunity to monetize otherwise costly treatment processes, and improve consumer acceptance of 'farmed fish'. By integrating traditionally separate agricultural production systems, the development of an environmentally and economically sustainable aquaculture production model may be realized and accepted. However, to ensure the sustainable development of an integrated recirculating aquaculture-based system utilizing excess nutrients from seafood production, renewed engineering principles must be applied to develop sound system design guidelines for realistic productivity estimates and economic sustainability, all the while addressing food safety concerns.

The traditional wastewater treatment process train is a 3-stage process beginning with a primary solids separation and removal process followed by a secondary biological treatment system for the purpose of biochemical oxygen demand (BOD) and preliminary inorganic dissolved nutrient reduction. The final, tertiary treatment process follows the biological treatment for the purpose of further nutrient reduction and possible disinfection. The size and scope of the treatment process in each stage is largely dictated by the effectiveness of each preceding treatment process. In other words, improving the efficiency in the first, particulate effluent capture stage would have the greatest impact on the subsequent treatment stages and reduce the size of the effluent treatment system overall. However, before the RAS particulate effluent stream may be utilized, a complete characterization in terms of other agricultural production needs must be performed to better understand how the waste may be utilized in another system as a resource.

All agricultural production systems produce a waste stream in one form or another. These waste products represent the degree of inefficiency of the production system in terms of output production as a function of input utilization. Traditional terrestrial animal production facilities irrigate the effluent on fallow grazing land. However, recirculating aquaculture facilities do not have the same land for irrigation; One benefit RAS has over terrestrial animal protein production systems is a significant reduction in land use. The integration of recirculating aquaculture with other production systems provides an alternative option to traditional and costly effluent treatment for sustainable and localized production of seafood and other agricultural products as a means for monetizing the effluent treatment process. However, the lack of research focusing on the fundamentals of engineering (e.g. *plant-available* nutrient production, efficient effluent nutrient *reuse*, system effects on integrated system *production rates*, *food safety*-centric system designs, *constant*-feed compost modeling, etc.) for the integration of two agricultural production systems has resulted in a series of trial and error attempts with varying degrees of success due to an incomplete understanding of the different system needs as they relate to one another. Systems-based research is needed to better support the development and integration of RAS with other agricultural production systems for optimized productivity, economic sustainability, and improved food safety.

The Agricultural Engineering research program at the University of New Hampshire in Durham, New Hampshire is focused on integrated agricultural production system research for the purpose of improving environmental and economic sustainability. The research is currently focusing on the utilization of the effluent from RAS for the purpose of monetizing the traditionally costly effluent treatment process prior to discharge. The elements of research include the characterization of the nutrient production (organic and inorganic) in terms of the needs for a separate agricultural production system.

This presentation will discuss the means by which a RAS effluent stream is characterized and managed, present potential integrated agricultural approaches for utilizing RAS wastes, and some methods which may be employed for characterizing the particulate effluent stream to enable an engineered integrated agricultural production system.

Fate of feed: Carbon, Nitrogen & Phosphorus cycles in near zero discharge RAS

Yogev, U.[1], Gross A.[1] *

[1]Zuckerberg Institute for Water Research, , Blaustein Institutes for Desert Research, Ben Gurion University of the Negev, Israel.

*Corresponding author: Amit Gross, Dept. of Environmental Hydrology & Microbiology, Zuckerberg Institute for Water Research (ZIWR), Blaustein Institutes for Desert Research (BIDR), Ben Gurion University of the Negev, Sede Boqer Campus, Midreshet Ben Gurion 84990 Israel.

Tel: 972 (0)8 6596896 Fax: 972 (0)8 6596909 Email: amgross@bgu.ac.il.

ABSTRACT

In response to increasing demand for aquaculture products and strict new regulations on organic matter and nitrogen discharge, inland closed recirculating aquaculture systems (RASs) are being developed as a viable eco-sustainable alternative to traditional aquaculture (e.g. pond, raceways and cage) because of their minimal environmental impact and controlled operation.

Feed is virtually the only source of solids in the RAS. It is estimated that only 20-30% of the feed is assimilated by the fish, while the rest is released to the water as feces or in a dissolved form. A basic RAS consists of a unit for removal of suspended solid (feces and uneaten feed- sludge) and a nitrification biofilter for oxidation of toxic ammonia to less toxic nitrate (NO_3^-). Typically, controlling nitrate in a non-toxic level is performed by water exchange, which might cause environmental pollution. Furthermore, in these systems alkalinity should often be added to balance pH. Advanced systems include a denitrification unit that reduces nitrate to dinitrogen gas. Consequently, water exchange is reduced as well as partial recovery of alkalinity that is achieved in the denitrification process.

Two additional potential pollutants from these systems that are receiving less attention are emission of nitrous oxide (N_2O) and excess of phosphorus that may end up in the environment. N_2O is a potent greenhouse gas which has a global warming potential 310 times that of CO_2 over a hundred year lifespan and also causes destruction of the ozone layer. Phosphorous is a limiting nutrients that in excess might cause eutrophication of water bodies. The goals of this study was to evaluate the fate of feed, specifically carbon, nitrogen and phosphorus by mass balance and dynamics (Kinetics experiments) in a novel near-zero discharge (<1% water exchange of system's volume per day) saline research scale RAS. It was hypothesized that understanding and optimizing these cycles would support an intense RAS without hampering water quality or fish performance, as well as decreasing its environmental footprint.

The research system was planned with the intent of growing 250-300 kg/m^3 and included a fish tank attached to three treatment loops: (a) a solid filter followed by an aerated nitrification fixed-film reactor; (b) a single-stage anoxic denitrification activated sludge bioreactor which utilizes fish sludge (backwash of solid filter) as a carbon source; (c) an anaerobic bio-digester (upflow anaerobic sludge blanket [UASB]) for treatment of excess denitrification biomass for the production of biogas.

Water, fish and sludge were sampled routinely from the different units for total suspended solid (TSS), total organic carbon (TOC), total nitrogen (TN), NO_3^-, NO_2^-, total ammonia nitrogen (TAN), total phosphorus (TP), PO_4^{-3}, pH, electrical conductivity (EC) and temperature. Mass balance of solids as well as carbon, nitrogen and phosphorus were performed. Evaluation of anammox activity in the denitrification reactor and N transformations in the RAS were described by a set of simple differential equations that were solved by differential equation solver in Polymath software (version 6.2).

It was demonstrated that the fish assimilated 37 ± 7% of applied feed (dry weight), while 51 ± 13% was accumulated in the fish sludge backwash from the solid filter. Fish sludge accounted for 39 ± 7%, 37 ± 2% and 75 ± 2% of carbon, nitrogen and phosphorus from applied feed, respectively. Phosphate and inorganic nitrogen concentrations in the fish tank were stable, averaging 20 mg/L and 42 mg/L respectively. In the biofilter, 32% of N and 10% of C (of the applied feed) were oxidized. The N was excreted by fish mainly as ammonia and then oxidized to nitrate, while C was oxidized to CO_2. In the denitrification reactor 44% of the feed- N and 11% of the feed- C were removed, mainly as N_2 and CO_2 respectively. In the UASB, 20% of feed- C was converted to biogas (55-62% CH_4), 2% was discharged in the effluent, and 4% accumulated as non-degradable solids in the reactor; From feed- N, 23% accumulated in the reactor due to precipitation as minerals and 4% was discharge. Nitrous oxide emissions from nitrification and denitrification were estimated as 0.6% and 0.8% respectively from feed- N.

It was noticed that phosphate concentration in the fish tank remained stable, although the system was continually fed by ~1% P. Current observations suggest that P was removed in the denitrification reactor likely by polyphosphate-accummulating bacteria, which were identified (e.g. *Gemmatimonas spp*). As water was transferred from the denitrification reactor to the UASB, these bacteria with excess P were also transported. In the UASB, the denitrification biomass was possibly degraded, enhancing phosphorus precipitation as struvite mineral, and consequently the phosphorus concentration in the RAS effluent was lower.

Diet-mediated floating salmonid faeces – benefits for RAS

Alexander Brinker, Mark Schumann

Fisheries Research Station, Argenweg 50/1, 88085 Langenargen, Germany
*Corresponding author: Alexander.Brinker@lazbw.bwl.de

ABSTRACT

Solid control is one of the main challenges in recirculating aquaculture systems (RAS). A novel approach to this problem is the application of functional feed incorporating an indigestible, low density additive to produce floating faeces. In recent lab trials low quantities of cork (Quercus suber) granules have resulted in consistent floating faeces, which can be quickly and easily removed from the system. Further trials in the commercial setting of a semi-recirculating trout farm, lead to marked improvements in solid removal efficiency, a halving of total ammonia nitrogen concentrations in production water from about 0.95 mg L–1 to 0.47 mg L–1, allowing a doubling of production for the same available water flow. Having proven the potential of floating faeces in facilitating nutrient load reduction and water efficiency, the current study applies the approach to RAS by comparing the waste removal in replicate fully closed recirculating systems over a full grow-out cycle. Rainbow trout of the test group were fed a conventional diet supplemented with 2.5 % cork granules (0.5-1 mm), resulting in buoyant faecal casts (density values below 1.00 g/m3). These were directly removed from the tank by a slotted overflow pipe at the water surface. The control group was fed an identical diet without the cork additive, and the resulting sinking faeces were removed by sedimentation in a classic manure pit. Wastewater from both treatments was cleaned mechanically by drum filtration and biologically by a moving bed bioreactor before being returned to the tanks.

While there was no apparent effect of cork supplementation on feed performance or mortality of rainbow trout, preliminary data show the average single-pass solid removal efficiency of cork-treated wastes was improved more than threefold (51 % vs. 16 %). In consequence, the biofilter performance was also increased; resulting significantly lowered TAN values in production water. TSS levels in the backwash stream of the drum filter were about five times greater in the cork diet treatment.

Floating faeces generating-diets show great promise as a powerful and cost-effective tool in the RAS environment, with the potential to improve solid removal efficacy well beyond what is currently possible with today's state-of-the-art mechanical treatment options.

Growth of Atlantic salmon fed alternative protein diets

Gary S. Burr[1*], William R. Wolters, Frederic T. Barrows, Ronald Hardy, Casey Grimm and Michael Dowd

[1]USDA, Agricultural Research Service, National Cold Water Marine Aquaculture Center, 25 Salmon Farm Road, Franklin, ME 04634

ABSTRACT

Improvements have been made with the development of cost-effective salmonid diets utilizing multiple ingredients to insure optimal growth and it is important to evaluate families involved in breeding programs with current diets and assess the gains from selective breeding and diet modifications. These studies evaluated the performance of Atlantic salmon fed diets containing alternative protein ingredients. The first study evaluated growth responses of Atlantic salmon fed three different diets; two fishmeal free diets and one fishmeal control diet. Salmon (mean weight 182.22 ± 1.70 g) were stocked into triplicate tanks for each dietary treatment. Fish were fed at 1.5% of their body weight for the first six weeks, 1.0% of their body weight for the next six weeks, and 0.75% of their body weight for the final 6 weeks. The fish in this study were grown on the experimental diets for 18 weeks to an average final weight of 591.97 ± 5.44g. There was a significant difference in growth between diets ($p = 0.0058$) with fish fed an animal/plant protein blend diet (mean weight gain 423.44 g) having significantly greater weight gain compared to fish fed the fishmeal reference diet (mean weight gain 395.47 g), but not significantly different than a poultry-by-product protein blend diet (mean weight gain 411.15 g, $p = 0.1396$). Results show that Atlantic salmon selected for growth on typical fish meal diet will also have superior performance on diets that contain alternative protein sources.

A second forty week feeding study was conducted with Atlantic salmon (*Salmo salar*) smolts in two recirculating aquaculture systems. Twelve salmon (average initial weight 117 g; initial density 9.4 kg/m^3) were stocked per tank. Two identical systems were used and contained either freshwater (0 ppt) or seawater (~30 ppt), with the temperature maintained at 13°C by heat pumps/chillers. Fish were fed one of two diets, a control diet containing fishmeal and fish oil, or a marine-free diet that did not contain fishmeal or fish oil. The diets were fed to 6 replicate tanks in each system. The feed conversion ratio [FCR] (1.14 to 1.29), SGR (0.63-0.66) or average weight gain (590-667g) did not differ significantly among dietary or water treatments. There was however, a significant diet × water interaction for average weight gain and fish fed the reference diet growing slightly better in seawater and fish fed the marine free diet growing slightly better in freshwater. The fish fed the marine-free diet had significantly lower whole body crude protein levels (54.4% to 59.4%) compared to salmon fed the reference diet. There were no significant differences in whole body crude lipid (9.3% - 10.8%), moisture (66.9% - 68.7%), or ash (2.6% - 3.1%) contents. Diet and water salinity did not affect the texture of the fillets, but fish grown in seawater had significantly higher color scores compared to fish grown in freshwater (8.4 a* - 6.1 a*). Fillet yield was not significantly affected by diet. The diet did affect the fatty acid composition of the fillets. Fish fed the marine-free diet had significantly less eicosapentaenoic acid (20:5 n-3) (1.6% of lipids – 9.9% of lipids) docosahexaenoic acid (22:5 n-3) (0.5% of lipids – 3.2% of lipids) compared to fish fed the reference diet. The salinity of the water did not affect fatty acid composition of the fillets. Overall the salmon had similar performance regardless of diet or water salinity, but did appear to have more pigment retained in the flesh when grown in seawater.

The final study consisted of two 16-week feeding trials were conducted with Atlantic salmon smolts to determine the effects of adding either 15%, 20% or 30% corn protein concentrate to the diet. The fish fed the diets containing the corn protein concentrates had similar growth and feed conversion (turning food into muscle mass) compared to fish fed a fishmeal reference diet. The protein, fat and moisture levels of the fish did not change with dietary treatment. On the second study, the fish fed the corn protein concentrate had fillets that had more yellow color compared to the fillets of fish fed reference diet, however the red pigment normally found in Atlantic salmon diets was not added to these diets. The effects of the corn pigments on a regular salmon diet needs to be assessed.

Effects of Feeding a Fishmeal-free Vs. a Fishmeal-based Diet on Post-smolt Atlantic Salmon *Salmo salar* Performance and Product Quality & Water Quality and Waste Production Rates in Recirculation Aquaculture Systems

John Davidson*[1], Frederic Barrows [2], Brett Kenney [3], Christopher Good [1], Karen Schroyer [1], and Steven Summerfelt [1]
[1] The Conservation Fund's Freshwater Institute, 1098 Turner Rd., Shepherdstown, WV 25443
[2] U.S. Department of Agriculture, Agricultural Research Service, Fish Technology Center, Bozeman, MT 59715
[3] West Virginia University, Div. of Animal and Nutritional Sciences, Morgantown, WV, 26506
* Presenting author's contact information: j.davidson@freshwaterinstitute.org
PH: 304-870-2221; FAX: 304-870-2208

ABSTRACT

The Atlantic salmon farming industry has been progressively decreasing the proportion of fishmeal used in commercial diets due to rising costs and sustainability concerns. A variety of alternate proteins have been identified to partially replace fishmeal in Atlantic salmon diets; however, complete fishmeal substitution has only recently been attempted by a limited number of commercial feed suppliers and is still under development. In addition, very little published research has described the effect of feeding alternate protein, fishmeal-free diets to Atlantic salmon. Coinciding with the trend to reduce or eliminate fishmeal in Atlantic salmon diets is a surge to design and evaluate culture system technologies that could enable increased production and expansion of this industry sector to meet rising consumer demand for salmon. Hence, recirculation aquaculture systems (RAS) are being evaluated and adopted as a farming technology for land-based smolt production, and in some cases full cycle production of market-size (4-5 kg) Atlantic salmon. Alternate protein diet formulations purposed for feeding Atlantic salmon cultured in RAS should meet the nutritional and biological requirements of the fish, but should also be compatible with the production system. Specifically, these diets should: 1) minimize the amount of nitrogen and phosphorous waste excreted by fish, 2) produce intact fecal matter that can be effectively captured and removed from the recycled flow, 3) result in water quality that is suitable for optimal growth, health, and welfare of cultured species, and 4) create waste production rates and corresponding effluent flows that allow compliance with local discharge standards. Alternate protein, fishmeal-free diets should also be capable of producing optimal product quality of market-size Atlantic salmon, including maintenance of the health benefits (omega-3 fatty acids) that consumers expect from this premium protein source.

In light of developing trends, a 6-month study was conducted to compare the effects of feeding a fishmeal-free (FMF) versus a traditional fishmeal-based diet (FM) on post-smolt Atlantic salmon performance, health, and welfare; as well as water quality and waste production rates of replicated recirculation aquaculture systems. At the conclusion of the study, product quality of salmon fillets was also assessed. Six identical RAS were stocked with 220 post-smolt Atlantic salmon (281 ± 5 g). Each experimental diet was fed to salmon in triplicate RAS. The primary protein ingredients used in the FMF diet were mixed nut meal, wheat flour, corn protein concentrate, and poultry meal. The FM diet was formulated to represent a commercial-type Atlantic salmon diet containing menhaden meal, poultry meal, soy protein concentrate, and blood meal. Fish oil, derived from whiting fish trimming waste, was used in the FMF diet to create a wild fisheries-in to farmed fish-out ratio of 0:1 as defined by Monterey Bay Aquarium's scoring criteria; while menhaden oil was the primary lipid used in the FM diet. Both diets were formulated with an approximate protein/fat ratio of 42/27. Feed was delivered via automated feeders during 24 equally-spaced events around-the-clock and a 24-h light photoperiod was utilized. Each experimental RAS was operated with relatively low water exchange, replacing enough water to maintain mean nitrate-nitrogen concentrations at <75 mg/L. System hydraulic retention times averaged 20 days (5 % of the system volume flushed daily), and average feed loading rate was 3.2 kg feed/daily make-up water volume.

Atlantic salmon growth rates, survival, and feed conversion ratios (FCR) were statistically similar ($P > 0.05$) between diets. Atlantic salmon fed the FMF and FM diets were 1.716 ± 0.076 and 1.720 ± 0.065 kg; cumulative survival was 99.8 ± 0.2 and 99.7 ± 0.3 %; and average FCR was 0.89 ± 0.03 and 0.90 ± 0.02, respectively. Significant differences in intestinal inflammation and blood chemistry were not detected between dietary treatments. The most notable water quality difference between treatments was for total phosphorous which was approximately four times greater in the culture water of RAS associated with the FMF diet, 4.2 ± 0.1 mg/L compared to 0.9 ± 0.0 mg/L for the FM diet. The FMF diet also resulted in greater total phosphorous (TP), carbonaceous biochemical oxygen (cBOD), and total suspended solids (TSS) demand per kg feed in the effluent ($P < 0.05$). The FMF and FM diets produced 0.009 ± 0.001 vs. 0.006 ± 0.001 kg TP/ kg feed; 0.079 ± 0.005 vs. 0.056 ± 0.005 kg cBOD/ kg feed; and 0.297 ± 0.028 vs. 0.221 ± 0.032 kg TSS/ kg feed, respectively. A higher percentage (52 %) of TSS was captured by radial flow settlers of RAS receiving the FMF diet compared to capture (35 %) by settlers associated with the FM diet suggesting that the FMF diet produced fecal matter with better settling characteristics. No differences were detected between dietary treatments for processing and fillet yields, fillet proximate composition, cook yield, fillet texture, or fillet color. The fillet fatty acid profile was also statistically similar between diets and important omega-3 fatty acids (EPA and DHA) were maintained in fillets.

This was the first research attempt to formulate a fishmeal-free diet for Atlantic salmon with this ingredient profile and one of few studies to demonstrate uncompromised Atlantic salmon performance and product quality when feeding a diet devoid of fishmeal. The fishmeal-free diet produced a significantly greater mass of three key wastes (total phosphorous, totals suspended solids, and carbonaceous biochemical oxygen demand); but culture system water quality was not visibly impacted by this difference, and fish performance was unaffected. The elevated phosphorous levels in the effluent resulting from feeding the fishmeal-free diet could be problematic depending on waste discharge requirements, but could also be perceived as advantageous for aquaponics applications. Albeit, the authors are confident that phosphorus excreta can be substantially reduced (if necessary) by lowering the inclusion level of dicalcium phosphate in the diet formulation. The successful use of mixed nut meal as an alternate protein to fishmeal adds to the diversity of ingredients available to feed suppliers and Atlantic salmon producers interested in using sustainable diets. The results from this study can be used by nutritionists to refine future formulations and by engineers and operators of RAS intending to use similar diets.

Recirculation feed for Atlantic salmon (*Salmo salar*)

Mikkel Detz Jensen [a] & Kim S. Ekmann [a]

[a] BioMar A/S, Mylius Erichsensvej 35, DK-7330 Brande, Denmark. E-mail: mdj@biomar.dk

Introduction

The RAS production of salmon smolts in both Norway and Chile has been steeply increasing for more than a decade. This has been caused partly by an increase in number of salmon smolts produced, but also by a trend of growing smolts bigger (up to 400 grams or more) prior to sea cage stocking.

As a consequence there is a demand for developing bespoke salmon smolt feeds for RAS with focus on:

- Superior growth performance (SGR)
- Improved feed conversion ratio (FCR)
- High nutrient digestibility and retention
- Stabilizing faecal matter

-in order to safely maintain high stocking densities and fast production cycles and reduce discharge of both organic and nitrogenous waste to a minimum without jeopardizing overall feed performance. Stabilization of fecal matter will not be treated in this presentation.

Methods

From a number of preceding trials, one experimental RAS diet was chosen to be tested against two commercially available diets for Atlantic salmon smolt in a growth trial in triplicates using 1m^3 tanks. Both commercial diets were known to have been previously used for salmon smolt production in RAS. The three diets were fed to salmon smolt growing from 20 – 175 grams over two time periods with a combined duration of 118 days. Fish were fed *ad libitum* for 24 hours per day at a temperature of 14°C. Light regime was 24L:0D throughout the trial.

Results

Fish fed the experimental RAS diet showed significantly lower FCR values (0.75) compared to the commercial diets (both 0.78), while no significant differences were found in SGRs between dietary treatments (2.06 – 2.11%d^{-1}). Protein and lipid digestibilities of the two commercial diets (88.7-90.3% and 95.9-97.2%, respectively) were significantly lower than observed in the experimental RAS diet (91.7% and 98.0%, respectively). Similarly, fish fed the experimental RAS diet displayed an improved retention of digested protein (55.2%) compared to fish fed the two commercial diets (51.9 – 52.5%).

Discussion

Collectively, these dietary measures allowed a reduction of nitrogen excreted *via* faeces and gills/urine of 20.5-33.6% and 10.6-13.1% per kg produced fish, respectively, and a reduction of dry matter feces excretion between 1.9-6.1% when using the proposed recirculation diet.

The presentation will also touch on the latest experience with reduced fish meal levels for smolts and RAS diets for larger salmon.

Modern Norwegian salmon diets, resource utilization and fish welfare challenges

Mari Moren (presenting author), Trine Ytrestøyl, Bente Ruyter

Nofima AS, P.O. Box 6122 Langnes, NO-9291 Tromsø, Norway

ABSTRACT

The worldwide salmon production in 2014 was 2 226 000 tons of which 1 202 000 tons was produced in Norway. In comparison, the Norwegian salmon production in 1994 was approximately 300 000 tons. Already at that time, most of the fish meal and oil produced in Norway went to fish feed (Waagbø et al 2001). During the same period, the feed factor (kg feed/kg fish) has not changed much, but has been relatively stable at 1.2, despite the implementation of selective breeding programs.

The increase in salmon production the last two decades has been possible due to the use of vegetable feed ingredients, replacing the fish meal and oil, which were the main protein and lipid sources 20 years ago. Today's Norwegian salmon diets consists of more than 70% vegetable ingredients. The dominating components of Norwegian salmon diets are soy protein concentrate and rapeseed oil. Other vegetable ingredients also used are wheat gluten, sunflower meal, pea protein concentrate, faba beans and corn as protein sources and wheat, pea and tapioca as binders (Ytrestøyl et al 2015). Increasing the content of plant ingredients in the diet result in a higher content of carbohydrates and fiber that are not utilized by the fish, and may thus increase the waste output from fish farming. This may have implications particularly in RAS and in freshwater lake aquaculture.

The limited availability of fish oil represents a major challenge. Eicosapentaenoic acid (EPA) and docosahexaenoic acid (DHA) are poly-unsaturated fatty acids that are only found in the marine food chain. These are essential nutrients and there are no viable alternative sources on the market. In 2000, 31% of the salmon diet was fish oil, while the diets today often contains less than 10%. Average amount of EPA and DHA in commercial diets in 2012 was 1.1% and 1.5%, respectively (Ytrestøyl et al 2015). One major question has been; what is the minimum requirement of EPA and DHA for salmon in order to maintain good growth and welfare? In an ongoing project, salmon was fed as little as 0.2% EPA and DHA from 400 to 3500 gram fish size. There were no differences in growth compared to the control fed 1.7% EPA and DHA. Classic requirement studies use growth as the indicator, so this could be interpreted that 0.2% was sufficient.

However, the mortality after handling stress (removal of sea lice) was almost 4 times higher in the group fed low EPA and DHA levels compared to the control (Ruyter et al. 2016). The surviving fish in the groups fed ≤1% EPA+DHA in the diet had also some signs of poor welfare, and had less astaxanthin in the filets. The color of the filets is one of the main quality parameters.

Another effect of replacing the marine ingredients with vegetables is the changes in the content of micronutrients and undesirables such as pollutants and herbicides. While some micronutrients, such as phosphorus become less available, the level of organic pollutants decrease. Residues of herbicides have become a new possible concern.

The search for new sources of EPA and DHA, in addition to fish oils, are ongoing, and some promising future sources are marine microalgae and genetically modified plant oil seeds . One of the obstacles so far is that these alternatives do not contain both EPA and DHA, but have either high levels of EPA or DHA .

To conclude, without vegetable replacements in the diets, the aquaculture industry could not have become as large as it is today. Even though aquaculture is considered an effective way to produce food for humans, there is a potential for improvements. This presentation will provide an overview of the main changes in Norwegian salmon diets during the last 20 years and comment on the challenges this gives in regarding resource utilization as well as fish welfare with emphasis on the data Nofima's researchers have obtained in recent projects.

References:

Ruyter, B., Bou M., Bæverfjord, G., Østbye, T.K., Ytrestøyl, T., Bjerke, M., Sigholt, T. og Berge, G. (2016) Langtidseffekter av lave omega-3 nivåer i fôr på laksens helse; FHF rapport prosjekt 900957 (project report in Norwegian).

Waagbø, R., Torrisen, O.J., Austreng, E. (2001) Fôr og fôrmidler – den største utfordingen for vekst i norsk havbruk. Utredning for Norges forskningsråd (Norwegian report for the Research Council of Norway).

Ytrestøyl, T., Aas T.S., Åsgård, T. (2015) Utilisation of feed resources in production of Atlantic salmon (*Salmo salar*) in Norway. Aquaculture; 448, 365-374

Efficient ozone transfer into RAS water and an innovative way to determine delivered ozone in site

Aikaterini Spiliotopoulou[1,2], Richard Martin[3], Henrik R. Andersen[1]

[1]Department of Environmental Engineering, Technical University of Denmark, Miljøvej 113, 2800 Kongens Lyngby, Denmark
[2] OxyGuard International A/S, Farum Gydevej 64, 3520 Farum, Denmark
[3] Water ApS, Farum Gydevej 64, 3520 Farum, Denmark

ABSTRACT

In recirculating aquaculture systems (RAS), the water quality changes continuously due to fish feed, excretions and makeup water or piping system, affecting system's equilibrium. Accumulation of organic and inorganic compounds, where proteins, ammonia and heavy metals are the most pronounced; creates toxic conditions for aquatic organisms, leading to system failure. The dissolved organic matter (DOM) varies among the different water sources, affecting the reaction rate of ozone and consequently its lifetime.

Ozone is a strong oxidizing agent, reacting rapid and in low concentrations, first with the easily degradable DOC (Eq.1) and inorganic pollutants, and then with the decreasingly reductive pollutants. If more ozone is dosed than the immediate demand by reducing pollutants, it will be decomposed to hydroxyl radicals (Eq.2), which are non-selective, highly reactive species oxidizing a range of recalcitrant dissolved pollutants (Eq.3).

$$DOC + O_3 \rightarrow DOC_{selectively\ oxidized} \quad \text{Eq. 1}$$

$$H_2O + O_3 \rightarrow O_2 + 2\,HO^{\bullet} \quad \text{Eq. 2}$$

$$DOC + HO^{\bullet} \rightarrow DOC_{radical\ oxidized} \quad \text{Eq. 3}$$

When ozone is introduced into water, bacteria load and dissolved organic matter (DOM) are diminished while redox level, water clarity and UV transparency are increased. Protein degradation is accelerated and coagulation, filtration and nitrification processes are improved. However, in a non-meticulously designed system, residual ozone (due to overdose) with longer lifetime will reach the culture tanks causing significant harm to cultured specie. Ozone has been reported to be toxic to a wide range of marine and freshwater organisms at residual concentrations between 0.01 mg/L and 0.1 mg/L. The risk to lose fish and the high ozonation cost are limiting parameters and contribute to a reluctance by the aquaculture industry to use ozone. Therefore, ozone should be properly delivered, efficiently dissolved and accurately controlled to ensure that it is completely consumed before returning to culture tanks.

The present study investigates the optimal technology to transfer ozone into water based on physicochemical model applied to different established delivery methods e.g. gas cone,

gravitation bubble column or venturi injector. Depending on the water quality (DOC, salinity, pH, temperature, etc.), which will be analyzed in advance in the laboratory, the three dissolving alternatives will be tested in site. Based on the water flow and the disinfection needs of the facility, it will be suggested which is the optimal gas transfer method. The transfer efficiency will be monitored with oxidation reduction potential (ORP) sensors in site and by a colorimetric assay which will be developed in the laboratory.
Water samples were collected and transferred to the laboratory for further analysis. Ozone measurement in water is usually achieved by a spectrophotometer utilizing a colorimetric assay, since ORP sensors do not determine it successfully. Therefore, the possibility to determine the delivered ozone dose by utilizing the natural fluorescence caused by certain proteins, which are contained into RAS is investigated. Preliminary experiments to test this hypothesis have been conducted in wastewater effluent providing satisfactory results. Since the aquaculture water is enriched with proteins it is expected that the fluorescence effect will be greater leading to an innovative ozone determination technology. The method is evaluated by comparing it with a colorimetric assay.

Removal of Fine Suspended Solids in Recirculating Aquaculture Systems using Fixed Bed Contact Filters

Lionel Hayter[1] and Michel F. Couturier[2]

[1]Sorensen Engineering Ltd., St. Andrews, NB, Canada (lionel@soreng.ca)
[2]Chemical Engineering Department, University of New Brunswick, Fredericton, NB, Canada (cout@unb.ca)

ABSTRACT

The buildup of fine solids in intensive recirculating aquaculture systems (RAS) can stress fish and cause adverse health effects such as gill irritation and increased susceptibility to disease. Maintaining low concentrations of fine solids within the recirculation loop is therefore a priority to ensure healthy fish. A high-void, fixed bed static filter, known as the contact filter, has shown potential for the removal of fine solids. The bed consists of bio-carrier media with a high porosity (77-78%) trapped between two porous support plates. Two different bio-carrier media (RK and Curler Advanced x-1) were investigated in this study. Tests were performed on full-scale contact filters in operation at salmon-smolt RAS and on a pilot-scale unit. The impact of fluid velocity (0.59-1.02 cm/s), bed depth (0.20 and 0.48 m), media type, and backwashing frequency on the overall and fractional solids capture efficiencies of the contact filter were assessed. The ability of the filter to remove ammonia was also determined.

The removal efficiency of 8.1-20 □m particles in the pilot-scale contact filter was initially minimal when using either media but increased to 39□6% after 11 days due to the development of a biofilm. The contact filter not only captured fine particles but also behaved as a bio-reactor with heterotrophic bacteria within the unit consuming the majority of the captured solids to produce carbon dioxide and biomass. In the process, the bacteria removed 14□5% of the incoming ammonia. Solids capture efficiency increased from day 1 to 11 as biofilm formed on the bio-carrier media and then remained relatively constant to the end of the monitoring period (24 days) indicating backwashing should be no more than once or twice per month. Detachment/re-entrainment was observed in mature contact filters but the majority of the detached particles were sloughed biofilm. Increasing fluid velocity resulted in reduced capture efficiency for all particle sizes. Conversely, increasing bed depth resulted in more efficient solids capture as well as prolonged effective filtration periods. No difference in performance was observed between the two bio-carrier media tested.

Cleaning of the contact filters was performed by (1) dropping the level of water in the bed below the top support plate, (2) injecting air to agitate the bed and dislodge the solids, and (3) draining the dirty water containing the dislodged particles. Air agitation for 30 minutes was found to dislodge 83±5% of solids from the media and re-deposition of the solids onto the media during draining appeared to be negligible. Nonetheless, a pulse of dirty water which temporarily deteriorated water quality was experienced when the contact filters were returned to service after cleaning. It is believed that solids remaining on the filter bottom and inside the inlet pipe after draining were the most likely cause of the pulse of dirty water.

Biofiltration Capacity Of A Moving Bed Biofilter In Sequence With A Submerged Floating Bead Bed Bioclarifier – Application To Water Recycling In Aquaculture System (RAS)

Timothy J. Pfeiffer, PhD[a], Rhine Perrin[a], Ronald F. Malone[a], PhD, P.E

[a]Aquaculture Systems Technologies, 108 Industrial Avenue, New Orleans, LA
tim.pfeiffer@astfilters.com

ABSTRACT

Biofiltration is a key process in recirculating aquaculture systems (RAS). The microbial oxidation of ammonia to nitrate is universal and vital to ensure that ammonia and nitrite do not accumulate in the fish culture system. In moving bed biofilters (MBBF) and submerged floating bead bed bioclarifiers, nitrifying bacteria colonization occurs as a fixed film on the media surfaces of each respective unit. This aerobic nitrification process is well described and numerous biofilter designs have been developed, tested, and reported. The numerous biofilter solutions are often site specific, based on operator preference, or country traditions. An ideal biofilter for all purposes does not exist and all types of biofilters have their pros and cons. Much effort and research has been expended in developing guidelines for reporting biofilter performance and optimization. This presentation describes the performance limitations and application of a common RAS unit operation for solids removal and biofilitration, namely a bead filter in sequence with a moving bed biofilter.

It has been observed that moving bed biofilters placed behind bead filters with floating bead media (polygeyser, propeller-wash, or bubble-bead) are often flow limited in their ability to convert ammonia. That is, the MBBR ability to convert total ammonia nitrogen (TAN) to nitrate is limited by the transport of TAN into the unit. The bead filter must be sized to provide the necessary flow. Consideration of the need for flow and recognizing the TAN conversion capabilities of the bead filter leads to the sizing approach described under the assumption nitrogen is limiting. Commonly assumed bead filter and MBBR ammonia conversion data is used to model several scenarios of varying TAN concentrations and system volumes to describe the need to consider water flow in the treatment sequence. The STELLA (v 9.1) from ISEE Systems was used to develop and evaluated the marrying sequence scenarios.

AUTHOR INDEX

RAStech Conferences
PO Box 11736
Blacksburg, VA
24062

ISBN 978-1-7138-1010-0

9 781713 810100

CURRAN ASSOCIATES INC.
proceedings.com

Integrating Monitoring and Modelling for Understanding, Predicting and Managing Sediment Dynamics

ICCE Symposium 2016

Proceedings of the International Association of Hydrological Sciences (PIAHS) Volume 375, 2017

Okehampton, United Kingdom
11 – 15 July 2017

IAHS
AISH